Journal of Interdisciplinary Science Topics
Volume 3

Contributions from:
University of Leicester Interdisciplinary Science Students
and
McMaster University Integrated Science Students

January-April 2014, University of Leicester, UK

EDITOR
Dr Cheryl Hurkett

Copyright © 2014 by Centre of Interdisciplinary Science, University of Leicester

All rights reserved. This book or any portion thereof may not be reproduced or used in any manner whatsoever without the express written permission of the publisher except for the use of brief quotations in a book review or scholarly journal.

First Printing: 2014
ISBN 978-1-291-89896-5
Centre of Interdisciplinary Science
University of Leicester
University Road
Leicester
LE1 7RH

www.le.ac.uk/iscience

Foreword

THE MODULE:

The *Journal of Interdisciplinary Science Topics* (JIST) forms part of the 10-credit 'Science in Content' module in the third year of both the BSc and Natural Science degrees. The module as a whole is intended to provide our students with experience of communicating information from the cutting-edge of scientific research. This is approached in two complementary stages. First, they develop familiarity with the current scientific literature by presenting and discussing current articles in the style of a research group seminar. Second, they write their own original papers for a peer reviewed undergraduate journal, the Journal of Interdisciplinary Special Topics (JIST). Both activities provide valuable practice in the communication skills required for completion of their 3rd and 4th year individual research projects and in their future careers.

The undergraduate journal (JIST) in particular is intended to provide students with hands-on experience of, and insight into, the academic publishing process. The activity models the entire process from paper writing and submission, refereeing other students' papers, sitting on the editorial board that makes final decisions on the papers, to finally publishing in an online journal.

The authorship phase starts by identifying an interesting scientific question or problem, applying existing scientific knowledge in a novel context and formulating a concise (1-2 sides of A4) paper in response. This process encourages both creativity and a deeper understanding of the scientific concepts. Students are particularly encouraged to produce mathematical models or exemplar calculations where possible in order to support their conclusions. Students may work individually or collaborate in small groups. In these respects the journal differs from other undergraduate journals which publish more extended papers based on students' capstone projects.

Once a paper has been submitted it is sent out to other students on the module (with no connection to the paper) who act as referees. These students have an essential role as it is their duty to maintain the scientific standards of the journal by providing an independent check of the contents and ensuring that mistakes do not appear in any published papers. After critically reviewing a paper the referees are required to submit a brief written report to the editorial board containing: a short summary of the work; their critical analysis of its contents and any constructive suggestions for changes; and a recommendation on whether the paper should be accepted, rejected or sent back for further rewriting and review.

All students sit on the journal's editorial board multiple times and take on member, chair and secretarial roles. The editorial board is crucial to the running of the journal as it is the responsibility of the board to assign referees (taking into account the balance workloads across the year group), to consider referees reports and ultimately to decide whether a paper will be published. It is the job of the editorial board to provide guidance to authors and referees and to maintain the standards of the journal and in so doing they may overrule both referees and authors where necessary.

The entire process is managed via professional grade free software supplied by *Open Journal Systems*, providing students with experience of the type of interface and management systems they will encounter when submitting papers to high impact journals such as *Science* or *Nature*.

ORDER OF PAPERS:

The papers herein are listed in the order which they were accepted for publication throughout the module.

It should be noted that the papers within this journal are only the papers that were accepted for publication. Other papers were submitted over the course of the module but did not reach publication status before the module concluded and I would like to acknowledge the work that went into them.

PERSONAL NOTE:

This was the first year that students from our student exchange partner, *McMaster University* in Ontario, were invited to submit papers to the journal as part of their own studies; staff from both institutions agree that it was a resounding success. As far as we are aware this is the first time there has been an international collaboration in an undergraduate journal as part of an undergraduate degree. I would personally like to thank the *McMaster* students for their engagement and willingness to embrace the ethos of the journal.

As always I was extremely impressed by the creativity displayed in locating the seed ideas for papers and the skill, wit and scientific ability used to translate these ideas into finished papers from the *Leicester* students. The addition of the *McMaster* papers imposed an extra refereeing load on the Leicester students which they handled with dedication and academic diligence. You should all be very proud of your efforts!

This volume contains 24 papers with inspiration ranging from mythology (*Golden Fleece; The Monkey King's Somersault*) to popular culture (*The Viability of coming in like a Wrecking Ball*) to TV programs (*Powers of Poison: The Science Behind Sherlock; Breaking Bad: Gus Fring's Face Blown Off*) and even thought experiments inspired by current events (*The Winter Olympics on Enceladus*).

The papers from this issue are cited by Google Scholar and have been subject of a series of University of Leicester press releases. The first paper to be cited by the media was *Can we power a space ship?* (iO9; an Italian technology website called 'Tom's Hardware'), which drew the media's attention to the rest of the volume. Subsequently a host of news outlets, including Sky News and the Telegraph (and perhaps less flatteringly the Mail Online!) covered a broad selection of papers from this issue and helped to promote them internationally. Articles were seen as far and wide as France, Italy, America, China and Russia. *How many lies could Pinocchio tell before it became lethal?* and *Does Winnie the Pooh have a B12 Deficiency?* in particular went viral and were the subject of several radio interviews (BBC Wales, BBC Warwickshire, ABC Melbourne) and numerous media articles; they even got a mention on the BBC's *Have I Got News for You*!

Dr Cheryl Hurkett

Centre for Interdisciplinary Science

EDITOR:

Dr Cheryl Hurkett

CONTRIBUTING AUTHORS:

University of Leicester

Yannic Chen
Deven Fosberry
Sean Gilmore
Chuqiao Huang
Pratik Lakhotia
Steffan Llewellyn
David McDonagh
Radvile Soryte
Stephanie Taylor
Somaya Turk

McMaster University

Emma Butcher
Aaron Goldberg
Rebekah Ingram
Nicole Lindsay-Mosher
Katie Maloney
Kira Moor
Daim Sardar
Vincent So

SPECIAL THANKS GO TO:

Professor Derek Raine, *University of Leicester*
Assistant Professor Sarah Symons, *McMaster University*
Dr Robert Cockcroft, *McMaster University*
Andrew Colgoni, *McMaster University*

Journal of Interdisciplinary Science Topics 2014

Contents

Does Winnie the Pooh have a B12 Deficiency? 1

The Viability of coming in like a Wrecking Ball 3

A Wild Magcargo Appeared 5

Powering Disney's Frozen with a Carnot Refrigerator 7

Determining Distance of Object Size from a Photograph 10

The Viability of Throwing Giant Tortoises onto Mines 13

The Frog Prince Transformation 15

A Scandal in Belgravia … for whom? 17

Golden Fleece: A Heavy Task 19

Evaluating *The Core*: The prospect of geodes 21

The Winter Olympics on Enceladus 24

How many lies could Pinocchio tell before it became lethal? 27

BoRK or BT? An Analysis for Vayne Players in League of Legends 30

The Monkey King's Somersault 33

Breaking Bad: Gus Fring's Face Blown Off 35

Playing 'The Floor is Lava' in Real Life 37

Can we power a spaceship? 40

The Curious Case of the Glowing Bones 42

Katniss's Flaming Wedding Dress in Real Life 45

Is Dr Conner's Regenerative Transformation Possible? 48

Powers of Poison: The Science Behind Sherlock 51

The Big Fat Lie About Burning Fat 54

Calculating the Punching Power of "One-Punch" Mickey 57

Golden Fleece: An Ancient Sheep 59

Does Winnie the Pooh have a B12 Deficiency?

Steffan Llewellyn and David McDonagh
The Centre for Interdisciplinary Science, University of Leicester
14/02/2014

Abstract
Decades of research studying the unique behaviour of Winnie the Pooh has provided a strong indication that a honey-specific diet could be causing a vitamin deficiency. A review is here conducted in the changes observed in the bear, and the likely cause of this behaviour.

Introduction
In 1958 a bear was first observed in the Hundred Acre Wood displaying anthropomorphic characteristics, including a preferred attire of a red t-shirt and significant bipedalism, which over a number decades has grown as a household name as Winnie the Pooh, or simply Pooh bear, both for its ambiguous evolutionary origin and charming personality. Despite his joyous demeanour, physiological changes have been noted throughout its lifetime, likely linked to a honey-specific diet. This has given rise to concern that vitamin deficiencies, particularly in B12, are now putting his health at risk. Highlighted here are the symptoms identified throughout the extensive period of observation recorded from 1988 to 2002, how these indicate a B12 deficiency, and the implications of this for Winnie the Pooh's lifestyle.

Patterns of Behaviour and Symptoms
Issues concerning Pooh Bear were first acknowledged from an aberrant behaviour towards acquiring honey. In similar changes noted in *Vulpes vulpes* adapting to an urban environment, behaviour deviating from that known of the genus *Ursidae* has been observed. Examples of this include the invasion of *Oryctolagus cuniculus* burrows and the sporadic formation of temporary mutualistic relationships amongst *juvenile Sus scrofa domesticus*, and *Equus africanus asinus* in the pursuit of honey [1]. A preference for this diet is likely to cause health problems over a long period of time, in light of the absence of many key vitamins required for numerous functions in the body, in particular Vitamin A, Vitamin E, Vitamin K, Thiamin and Vitamin B12 (figure 1). Symptoms identified from the known records of Pooh Bear are summarised here to infer potential health problems.

Vitamins		
Amounts Per Selected Serving		%DV
Vitamin A	0.0 IU	0%
Vitamin C	1.7 mg	3%
Vitamin D	~	~
Vitamin E (Alpha Tocopherol)	0.0 mg	0%
Vitamin K	0.0 mcg	0%
Thiamin	0.0 mg	0%
Riboflavin	0.1 mg	8%
Niacin	0.4 mg	2%
Vitamin B6	0.1 mg	4%
Folate	6.8 mcg	2%
Vitamin B12	0.0 mcg	0%
Pantothenic Acid	0.2 mg	2%
Choline	7.5 mg	
Betaine	5.8 mg	

Figure 1: The vitamin nutritional information for a 339g serving of honey [2].

Firstly, a characteristic yellowing of the skin is present in comparing initial photographs of the animal and those taken at a later date (figure 2). A restricted gait is also observed in its apparent preferred bipedalism, which, while unusual for *Ursidae* to adopt for long periods, can be seen to show limited lower joint movement. Frequent memory loss and fatigue are further present, where the creature frequently forgets levels of honey reserves, encouraging reliance on the reserves of other species.

In light of these symptoms, of the vitamins highlighted, characteristics observed in Winnie the Pooh appear to correlate with a B12 deficiency. Such a condition is common in those with

Figure 2: A comparison between the skin tone of Winnie the Pooh in 1928 (left) [3] and the period between 1988-2002 (right) [4].

restricted diets, such as vegetarians and vegans, coinciding with anaemia where the patient is often tried, easily fatigued and shows a paling of the skin [5, 6].

Potential Treatment
As a natural B12 deficiency is rarely observed in animals, confirmation of this condition in Pooh Bear could give insight into the creature's evolutionary origins. While treatments traditionally used in *Homo sapiens* vary, it is recommended that initial cobalamin tests are first carried out to quickly assess whether this condition is found to be true. As well as ensuring the health of this childhood icon, such a study could also be of interest to the pharmaceutical industry aiming to bridge the gap between B12 deficiency treatments in *Homo sapiens* and those used in livestock [7].

Acknowledgements
We would like to thank Walt Disney and his research group for making their data available to the public, and Alan Alexander Milne for his pioneering work in first discovering this unique species of *Ursidae*.

References

[1] DISNEY, 1988, The New Adventures of Winnie the Pooh: Friend In Deed, Disney
[2] CONDE, N., 2013---last update, Nutrition Facts: Honey [Homepage of Nutrition Data], [Online]. Available: http://nutritiondata.self.com/facts/sweets/5568/22014].
[3] HARRIS, P., 2013. Found....the sketch that first captured magic of Poohsticks: Original drawing depicting Christopher Robin, Piglet and Pooh playing game emerges after 85 years. The Daily Mail.
[4] DISNEY, 2014---last update, Winnie the Pooh Wiki [Homepage of Disney], [Online]. Available: http://disney.wikia.com/wiki/Winnie_the_Pooh2014].
[5] ANNE---METTE, H. and EBBA, N., 2006. Diagnosis and treatment of vitamin B12 deficiency. An update. The Hematology Journal, 91(11), pp. 1506---1512.
[6] STABLER, S.P. and ALLEN, R.H., 2004. Vitamin B12 Deficiency as a Worldwide Problem. Annual Review of Nutrition, 24, pp. 229---326.
[7] GENESIS, n.d., Vitamin B12 Deficiency in Sheep and Cattle, Ancare, 0705, Available: www.ancare.com

Journal of Interdisciplinary Science Topics

The Viability of coming in like a Wrecking Ball

David McDonagh
The Centre for Interdisciplinary Science, University of Leicester
03/03/2014

Abstract
A pop song that made the charts in late 2013 alludes to the idea of entering a given location analogous to a wrecking ball, raising questions over if such a feat is possible. Perhaps more significantly, the singer claims to have impacted both love and ostensibly the walls of someone's house with similar momentum at some point, providing a somewhat unique case in studying the effects of shock on human beings. Both claims are investigated with a view to their viability, concluding that any human behaving like a wrecking ball would likely result in serious injury.

Introduction
Popular music has a rich history of musicians stating extraordinary feats, from The Beatles claiming an eight-day week, to Westlife achieving human-powered flight without wings. Such claims must follow the basic known principles of science if they are to be believed, and can often be investigated through making simple approximations. The Miley Cyrus song *Wrecking Ball* is no exception to this, and can be scrutinised using basic classical mechanics. Of particular focus here are the chorus lyrics "*I came in like a wrecking ball. I never hit so hard in love. All I wanted was to break your walls*".

I came in like a wrecking ball
Wrecking balls are typically on the order of 1000-7000 Kg [1, 2, 3], and are a common tool in industrial demolition, where a mass is attached to a rope and risen to a certain height by a crane, before being left to free fall under gravity. Such motion can be modelled as a simple pendulum, where the potential energy of the ball is converted into kinetic energy, assuming air resistance is negligible:

$$\frac{1}{2}mv^2 = mgh,$$

where m is the mass of the wrecking ball, v is the velocity, h is the height of the ball and g is the acceleration due to gravity. As the wrecking ball is moving in an arc rather than in free fall, h can be calculated using simple trigonometry, as shown in figure 1.

From this, h can be rewritten as

$$h = l - l\cos(\theta).$$

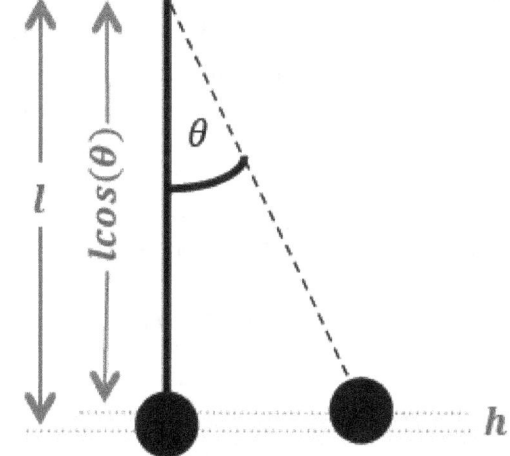

Figure 1: Identifying h in terms of the wrecking ball rope length and the angle the ball is raised. Drawn using MS Word.

In rearranging for velocity, the impact velocity can be obtained where the potential energy of the ball is fully converted into kinetic energy, given as

$$v = \sqrt{2gl(1 - \cos\theta)},$$

where l is the length of the rope of the crane, and θ is the angle raised before release. The momentum of the wrecking ball is then the product of the impact velocity and its mass. Taking modest values of a wrecking ball with a mass of 1000kg raised by a 10m rope to 60° gives a momentum of

$$P = mv$$

$$P = 1000 \times \sqrt{2 \times 9.81 \times 10(1 - \cos 60°)}$$

$$P = 9904.54 \; kg \; ms^{-1}.$$

Assuming now a mass of 70kg, the required velocity for the equivalent momentum for a human being is then

$$v = \frac{9904.54 \; kg \; ms^{-1}}{70 \; kg} = 141.49 \; ms^{-1},$$

or around 316 mph. As current theoretical limits to human speed are placed at around 40mph [4], these results make it unlikely that a human being could enter somewhere like a wrecking ball, without being propelled.

I never hit so hard in love/All I wanted was to break your walls

Upon making contact with a wall, a force is exerted on the object as it suddenly decelerates to zero velocity. This will be dependent on the time over which the force is applied, which, for concrete, steel and other common building materials for which wrecking balls are used, will be short due to their rigid structure. Estimating a value on the order of tens of milliseconds results in a deceleration of

$$a = \frac{0 \; ms^{-1} - 141.49 \; ms^{-1}}{0.05 \; s}$$
$$a = -2829.80 \; ms^{-2},$$

equivalent to around 288g, which, from Newton's second law of motion gives a force of magnitude 198086N. An assumption is made here in the object not breaking through or damaging the wall, as the song implies breaking through the walls was unsuccessful.

Discussion and Conclusion

Decelerations in the area of the value calculated are well beyond known limits to what a human being can stand without severe injury [5], hence it is unlikely that such a feat could be achieved under these conditions. As the damage inflicted by an impact is dependent on the time over which the force is applied, it could be possible to identify a material in which the deceleration is gradual enough to remain within the current safe limits of deceleration for human beings. However, it is the view of the author that wrecking balls would not be used on these types of materials and hence the analogy would no longer be consistent. Based on these findings, it is clear that a human being cannot possess the characteristics of a wrecking ball without sustaining significant injury, and other objects should be sought as an analogy.

References

[1] Hudgins, H.T, 1987, Demolition of Concrete Structures, Hudgins & Company, Inc.
[2] Rent---A---Crane, n.d, Bringing down the Cooling Tower, available at: http://www.rentacrane.com/demolition.htm
[3] Construction Equipment company, n.d, Demolition Wrecking Balls, available at http://www.constructionequipmentcompany.com/id34.html
[4] Peter G. Weyand, Rosalind F. Sandell, Danille Naomi Leoni Prime, and Matthew W. Bundle, 2010, The biological limits to running speed are imposed from the ground up. Journal of Applied Physiology
[5] Piersol, A.G and Paez, T.L, 2009, Harris' Shock and Vibration Handbook, McGraw Hill

A Wild Magcargo Appeared

Yannic Chen
The Centre for Interdisciplinary Science, University of Leicester
03/03/2014

Abstract
There are many Pokémon with unique characteristics that are not just impossible for a living organism but would have serious physical consequences. Magcargo is one such Pokémon that is being investigated in this paper. Its internal temperature of 18,000 °F would not only be extremely harmful to surrounding organisms, but a sudden appearance would most likely kill any person before they knew what had happened.

Introduction
Magcargo is a small (0.79 m) amorphous Lava Pokémon with the national Pokédex number 219. Its appearance is that of a snail, but with a bright red body composed of magma. Its shell is composed of grey, hardened magma that is very brittle and breaks upon the slightest touch. The internal temperature of Magcargo is roughly 18,000 °F that causes water to evaporate on contact. It can reform its body by dipping itself in magma [1]. The aim of this article is to investigate the effect of Magcargo and the effect on the average human if "A wild Magcargo appeared!" based on a YouTube video "3 Pokémon who could kill you" by VSauce [2].

Temperature
The internal temperature of Magcargo is 18,000 °F or 10,255 K, which is nearly twice as hot as the surface temperature of the sun (6,000K)[2].

There are 3 modes in which heat can be transferred: Conduction, convection and radiation. For simplicity, this paper models the outside temperature to be the same as the internal temperature. Assuming that Magcargo has a spherical body with a radius of 0.395 m gives surface area of 1.96 m^2. Its natural environment would mostly transfer heat through air convection and radiation (conduction does not apply to fluid like medium). Convectional heat dissipates very quickly in an open environment, such as the natural habitat of Magcargo. Thus, most of the heat transfer is through radiation which can be calculated using the Stefan-Boltzmann equation:

$$\frac{Q}{\Delta t} = \varepsilon \sigma A (T^4 - T_0^4),$$

where ε is the emissivity of the medium, A is the area, σ is the Stefan-Boltzmann constant ($5.67 \times 10^{-8} W\,m^{-2}\,K^{-4}$) [3], T is the source temperature and T_0 is the environmental temperature.

The lowest limit of emissivity of lava is 0.5 [4] and using the Stefan-Boltzmann equation this results in a transfer of 6.15×10^8 W to its surroundings (see below).

$$\frac{Q}{\Delta t} = (0.5)(\sigma)(1.96 m^2)(10{,}255 K^4 - 313.15 K^4)$$
$$= 6.145 \times 10^8 \, W$$

Assuming the specific heat capacity of air is 1.0 $kJ\,kg^{-1}\,K^{-1}$ and the density of air is 1.127 $kg\,m^{-3}$ [2], the temperature of air around Magcargo in a 10m radius (of an isolated system) would increase by:

$mass\ of\ air: density \times volume$
$$= (1.127\ kg\ m^{-3})\left(\frac{4}{3}\pi(10m)^3\right)$$
$$= 4{,}721\ kg$$

$Temperature\ change$:
$$\frac{6.145 \times 10^5\ kJ\ s^{-1}}{(1.0\ kJ\ kg^{-1}\ K^{-1})(4{,}721\ kg)} = 130\ Ks^{-1}$$

At the highest possible emissivity of lava (black body = 1) [4], the energy emission and temperature change will double. While heat in the atmosphere dissipates very quickly, the amount of heat that Magcargo radiates will nevertheless heat the surrounding air to relatively high temperatures.

Magcargo and the human body

If a Magcargo does appear suddenly at a distance of 10 m from the observer, then a lot of the energy would be absorbed by the human body. For a human with a height of 1.8 m and width of 0.5 m, the exposed surface area of the observer facing towards the Pokémon is 0.9 m^2. Since the rate of heat transfer is uniform in all directions from a spherical source, a sphere of heat surrounding the source would be produced. If the human is stood at a 10m distance, the area of this sphere would be:

$$4\pi(10m)^2 = 1256.6 \, m^2.$$

The human would take up 7.16×10^{-4} of the sphere's area as:

$$\frac{0.9 \, m^2}{1256.6 \, m^2} = 7.16 \times 10^{-4}$$

The proportional amount of heat transfer to the human would therefore be:

$$6.145 \times 10^8 W \times 7.16 \times 10^{-4} = 4.4 \times 10^5 \, W.$$

The average specific heat capacity of the human is 3.57 kJ kg^{-1} K^{-1} [5]. Assuming that the same human has a mass of 65kg, $4.4 \times 10^5 W$ would heat him up at a rate of:

$$440 \, kJ \, s^{-1} / (3.57 \, kJ \, kg^{-1} \, K^{-1} \times 65kg)$$
$$= 1.9 K s^{-1}.$$

With such a high temperature change, a human would die instantly. At highest emissivity, Magcargo would heat up the body twice as fast and blood at body temperature will start to boil within 17 seconds.

Conclusion

The magnitude of the hazard of Magcargo certainly should not be underestimated, but the video [2] overestimates in saying "anything within 50 metres would burst into flames", which would only be true for easily burnable objects standing at such distances for a long periods of time. It produces between 600 MW and 1.2 GW of power in through heat radiation that can heat the surrounding air by hundreds of degrees within a very short time. Magcargo travels very slowly and a heat gradient will be produced away from it, allowing its presence to be felt beyond hazardous distances before coming to close, which should give enough time to avoid confrontation. If a Magcargo does appear suddenly in front of you, at a far enough distance, the body might survive long enough to escape. The distance depends on the absorptivity and heat resistance of the body and is open for further investigation. Nevertheless, distances around 10m and closer would kill a normal human very quickly.

References

[1] BULBAPEDIA, no date. BULBAPEDIA. [Online] Available at:
http://bulbapedia.bulbagarden.net/wiki/Magcargo [Accessed 30 January 2014].
[2] 3 Pokémon Who Could Kill You, video, Vsauce3, 19 November 2013[Online] Available at:
http://www.youtube.com/watch?v=Lh-7kEk1ceE [Accessed 30 January 2014].
[3] Tipler, P. A. & Mosca, G., 2008. Physics for Scientist and Engineers. 6 ed. New York: W.H. Freeman and Company.
[4] Pinkerton, H., James, M. & Jones, A., 2002. Surface Temperature Measurements of Active Lava Flows on Kilauea Volcano, Hawaii. Journal of Vocanology and Geothermal Research, 113(1-2), pp. 159-176.
[5] Human Body and Specific Heat. (no date). The Engineering ToolBox. Accessed 31. January 2013 from
http://www.engineeringtoolbox.com/human-body-specific-heat-d_393.html

Journal of Interdisciplinary Science Topics

Powering Disney's Frozen with a Carnot Refrigerator

Aaron Goldberg
Honours Integrated Science Program, McMaster University
19/02/2014

Abstract
Frozen is Disney's latest film, in which the character Elsa unleashes winter on her entire kingdom. This paper quantifies the amount of water frozen and the amount of work required by a Carnot refrigerator to do so, arriving at values of $5.49772788 \times 10^{12}$ moles and 5.8×10^{15} Joules, respectively.

Introduction
Frozen is a 2013 award-winning film, featuring the Snow Queen Elsa, who has the ability to create snow and ice from thin air. In the film, she inadvertently freezes the fjord around the capital city Arendelle, plunging the entire kingdom into winter [1]. The present paper discusses the amount of energy a Carnot refrigerator would require to cause this freeze.

Total Amount of Water
The city of Arendelle was inspired by the Norwegian fjord Nærøyfjord (Figure 1) [2].

Figure 1: A comparison between the animated city of Arendelle, above, and Nærøyfjord, below [2].

Nærøyfjord is at least 18 km long and 500 m wide [3], and so its surface area is about 9×10^6 m². When, in the film, the ice begins to crack, it is apparent that about 1 m of ice is floating on top of the water. Using the density of ice as 0.9167 g mL⁻¹ [4] and of freshwater as 1.000 g mL⁻¹ [5], the total volume of ice is

$$Surface\ area \times \frac{1\ \text{m}}{1 - \dfrac{Ice\ density}{Freshwater\ density}} \quad (1)$$
$$= 108043217\ \text{m}^3.$$

Assuming this density of ice, the total mass of ice is 9.9043217×10^{13} g. With the molar mass of water being 18.0153 g mol⁻¹ [6], this is equivalent to $5.49772788 \times 10^{12}$ moles of ice.

Isobaric Molar Heat Capacity of Water
To calculate the molar enthalpy change of freezing this amount of water, one must first know how the molar heat capacity of water behaves through the associated temperature range. Heat capacity data for water [7] and ice [8] have been tabulated in Table 1.

Using Microsoft Excel to fit these data to a curve, the equations for isobaric heat capacities of water close to its melting point are

$$\bar{C}_{P,water}(T) = (0.00105345\ T^2 - 0.628043\ T + 168.952)\ \text{J mol}^{-1}\text{K}^{-1} \quad (2)$$

and

$$\bar{C}_{P,ice}(T) = (-0.000833751\ T^3 + 0.603612\ T^2 - 145.388\ T + 11677.7)\ \text{J mol}^{-1}\text{K}^{-1} \quad (3).$$

Table 1: Isobaric heat capacity versus temperature for water and ice at 1 atm pressure:

Water Temperature (K)	C_p (J mol^{-1} K^{-1})	Ice Temperature (K)	C_p (J mol^{-1} K^{-1})
273.16	76.014	230.08	25.389
275.16	75.898	236.19	25.527
277.16	75.799	242.40	28.430
279.16	75.713	249.31	28.882
281.16	75.640	256.17	27.577
283.16	75.577	262.81	26.372
285.16	75.523	267.77	18.682
287.16	75.476		
289.16	75.437		
291.16	75.404		
293.16	75.377		
295.16	75.354		
297.16	75.335		

Total Enthalpy Change

Assuming the water transitioned from an ambient 20°C to ice at −15°C, the enthalpy of reaction follows the formula

$$\Delta_{reaction}\bar{H} = \int_{293.15\,K}^{273.15\,K} \bar{C}_{P,water}(T)\, dT + \Delta_{fusion}\bar{H}° + \int_{273.15\,K}^{258.15\,K} \bar{C}_{P,ice}(T)\, dT. \quad (4)$$

This enthalpy is the sum of the enthalpy change in liquid water going from 20°C to 0°C, the enthalpy change of water at 0°C freezing into ice, and the enthalpy change of ice going from 0°C to −15°C. Evaluating (4) using (2), (3), and $\Delta_{fusion}\bar{H}° = -6008.224$ J mol^{-1} [8] leads to $\Delta_{reaction}\bar{H} = -7833.180$ J mol^{-1}. From the above result for the total amount of water, it follows that $\Delta_{reaction}H = -4.30646921 \times 10^{16}$ J.

Using a Carnot Refrigerator

It is well known that the most efficient heat engine is a Carnot engine, which harnesses the temperature difference between two reservoirs to do work. This can be done in reverse, harnessing work to drive a temperature difference between two reservoirs, and is known as a Carnot refrigerator. In this case, the latter two reservoirs are the ice at −15°C, and the air at 20°C. The coefficient of performance of the refrigerator is the ratio of the heat flow between the two reservoirs and the work input required for such, and follows

$$\frac{T_{cold}}{T_{hot} - T_{cold}} = 7.38. \quad (5)$$

By isolating for work, it can be found that the work required is equal to the total enthalpy change divided by the refrigerator's coefficient of performance. As such, the total mechanical work required by a Carnot refrigerator to power the transformation of the entire fjord from water to ice is 5.8×10^{15} Joules.

Conclusion

It has been shown that in *Frozen*, Elsa froze approximately 5.5×10^{12} moles of water. To accomplish Elsa's feat, a Carnot refrigerator would require 5.8×10^{15} Joules of energy. This amount is equivalent to the energy released by the Hiroshima nuclear bomb 115 times over, or that released by 63 Nagasaki nuclear bombs [9]. This immense number puts Elsa's power into perspective, implying either that the Snow Queen has enormous strength, or that Disney underestimated the ramifications of their animated fantasy.

References

[1] Walt Disney Animation Studios, *Frozen*, 2013.
[2] Dawson March, C., *Disney's Frozen: Inspired by Norway's beauty*, The Globe and Mail (2013). http://www.theglobeandmail.com/life/travel/disneys-frozen-inspired-by-norways-beauty/article15617632/
[3] *Nærøyfjorden*, Norwegian Encyclopedia (2013). http://snl.no/N%C3%A6r%C3%B8yfjorden
[4] Ginnings, D.C. and Corruccini, R.J., *An Improved Ice Calorimeter – the Determination of its Calibration Factor and the Density of Ice at 0°C*, Journal of Research of the National Bureau of Standards 38, 583-591 (1947). http://nvlpubs.nist.gov/nistpubs/jres/38/jresv38n6p583_A1b.pdf

[5] Barlow, P.M., *Ground Water in Freshwater-Saltwater Environments of the Atlantic Coast*, U.S. Geological Survey, 1262, 14 (2003). http://pubs.usgs.gov/circ/2003/circ1262/

[6] National Institue of Standards and Technology, *Water*, NIST Standard Reference Database 69: NIST Chemistry WebBook (2011). http://webbook.nist.gov/cgi/cbook.cgi?Source=2002WAG%2FPRU387-535

[7] National Institue of Standards and Technology, *Isobaric Properties for Water*, NIST Standard Reference Database 69: NIST Chemistry WebBook (2011). http://webbook.nist.gov/cgi/fluid.cgi?P=1&TLow=270&THigh=298.15&TInc=2&Digits=5&ID=C7732185&Action=Load&Type=IsoBar&TUnit=K&PUnit=atm&DUnit=mol%2Fl&HUnit=kJ%2Fmol&WUnit=m%2Fs&VisUnit=uPa*s&STUnit=N%2Fm&RefState=DEF

[8] Giauque, W.F. and Stout, J.W., *The Entropy of Water and the Third Law of Thermodynamics. The Heat Capacity of Ice from 15 to 273°K*, Journal of the American Chemical Society 58 1144–1150 (1936).

[9] Penney, W.G., Samuels, D.E.J., and Scorgie, G.C., *The Nuclear Explosive Yields at Hiroshima and Nagasaki*, Philosophical Transactions of the Royal Society of London. Series A, Mathematical and Physical Sciences 266, 357-424 (1970).

Determining Distance of Object Size from a Photograph

Chuqiao Huang
The Centre for Interdisciplinary Science, University of Leicester
19/02/2014

Abstract
The purpose of this paper is to create an equation relating the distance and width of an object in a photograph to a constant when the conditions under which the photograph was taken are known. These conditions are the sensor size of the camera and the resolution and focal length of the picture. The paper will highlight when such an equation would be an accurate prediction of reality, and will conclude with an example calculation.

Introduction
In addition to displaying visual information such as colour, texture, and shape, a photograph may provide size and distance information on subjects. The purpose of this paper is to develop such an equation from basic trigonometric principles when given certain parameters about the photo.

Picture Field of View
First, we will determine an equation for the horizontal field of view of a photograph when given the focal length and width of the sensor.

Consider the top down view of the camera system below (figure 1), consisting of an object (left), lens (centre), and sensor (right). Let d_1 be the distance of object to the lens, d_2 be the focal length, l_1 be the width of the object, l_2 be the width of the sensor, and α be the horizontal field of view of the photograph. Note that the object is wide and close enough such that its image, when in focus, takes up the entire width of the sensor.

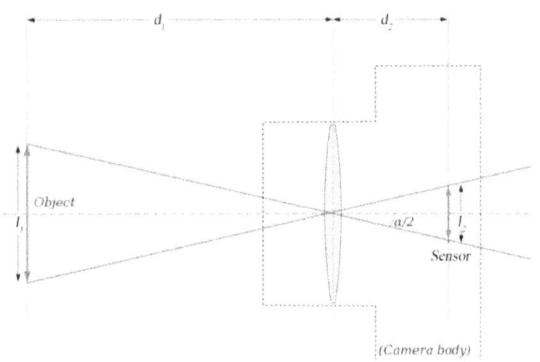

Figure 1 – A simple camera model [1].

On first glance, the relationship between the focal length (d_2) and the horizontal field of view (α) may be expressed as a tan (opposite/adjacent) relationship:

$$\tan 0.5\alpha = \frac{0.5 l_2}{d_2} \quad (1.0)$$
$$\alpha = 2\arctan \frac{0.5 l_2}{d_2} \quad (1.1)$$

Picture Angular Resolution
With the image dimensions, we can make an equation for one-dimensional angular resolution. Since l_2 describes sensor width and not height, we will calculate horizontal instead of vertical angular resolution:

$$\alpha_{hPixel} = (2\arctan \frac{0.5 l_2}{d_2}) \div n_{hPixel} \quad (2.0)$$
$$\alpha_{hPixel} = \frac{2}{n_{hPixel}} \arctan \frac{0.5 l_2}{d_2} \quad (2.1)$$

Here, α_{hPixel} is the average horizontal field of view for a single pixel, and n_{hPixel} is the number of pixels that make up a row of the picture.

Object Angular Diameter
If the width of the object on the image is measured, we can develop an equation for the horizontal angular diameter of the object:

$$\alpha_{hObject} = (\frac{2}{n_{hPixel}} \arctan \frac{0.5 l_2}{d_2}) \times n_{hpObject} \quad (3.0)$$
$$\alpha_{hObject} = \frac{2 n_{hpObject}}{n_{hPixel}} \arctan \frac{0.5 l_2}{d_2} \quad (3.1)$$

Here, $\alpha_{hObject}$ is the horizontal angular diameter of the object and $n_{hpObject}$ is the number of pixels in a row that make up the object.

For a sensor constructed such that the vertical and horizontal spacing between pixels are equal, the following formula may be used to measure the angular diameter in any direction:

$$\alpha_{Object} = \frac{2n_{pObject}}{n_{hPixel}} \arctan\frac{0.5l_2}{d_2} \quad (3.2)$$

Here, α_{Object} is the angular diameter, and $n_{pObject}$ is the number of pixels that make up the object in any direction.

Equation 3.0 and 3.2 assume that the angular resolution of a pixel is the same regardless of its position in the picture. In reality, this is impossible due to the planar nature of the camera sensor and the spherical nature of the field of view (figure 2). As camera lenses generally output images with rectilinear projections [2], equations 3.1 and 3.2 are most accurate when the object of interest is offset from the centre such that its constituent pixels have field of views close to the average (equation 2.1 output). The offset is dependent on the focal length.

Figure 2 – A rectilinear projection; each square represents the same field of view. Notice how the field of view of an individual pixel decreases towards the edges [3].

Object Distance / Length
An object with a set angular diameter may be an infinite combination of distances and sizes. For example, a further away large object will occupy the same angular diameter as a smaller object at closer distance. Glancing back to figure 1, this relationship may be approximated as:

$$\tan 0.5\alpha_{hObject} = \frac{0.5l_1}{d_1} \quad (4.0)$$
$$\alpha_{hObject} = 2\arctan\frac{0.5l_1}{d_1} \quad (4.1)$$

Combining equations 3.2 and 4.1, we form an equation relating object distance/diameter to the conditions under which a photograph was taken.

$$\arctan\frac{0.5l_1}{d_1} = \frac{n_{pObject}}{n_{hPixel}} \arctan\frac{0.5l_2}{d_2} \quad (4.2)$$

It is important to note that l_1 is in a single plane parallel to the plane of the sensor. Therefore, the equation will be most accurate when applied to an object such that, at all points it occupies in the picture, is at constant distance from the camera.

Example Calculation
For an example calculation, we will be calculating the distance of a Taiga Bean goose from the photographer. The photograph (figure 3) was taken with a Canon T1i, which has a sensor size of 22.3 x 14.9mm [4]. The picture itself has dimensions of 4752 x 3168px and was taken with a focal length of 85.0mm. The height of the goose in the picture is 1302px, and the average Taiga Bean goose has a height of 30.3cm [5]. We assume no distortion, and equal sensor horizontal and vertical pixel spacing.

Figure 3 – A pair of Taiga Bean geese crossing a road. The subject of our calculations is in front [6].

By adding the values into equation 4.2, we arrive at:

$$\arctan\frac{0.5 \times 303mm}{d_1} = \frac{1302px}{3168px} \arctan\frac{0.5 \times 14.9mm}{85.0mm}$$
$$d_1 = 4214.74mm$$

Thus, the goose was 4.21m from the photographer when the picture was taken.

Conclusion
Currently, how well the equation predicts object distance or size is unknown. A method of testing would involve taking a picture of an object where the distance and object size are both known, calculating both values using the equation, and comparing real values with calculated values.

Furthermore, the equation is limited to working with objects that remain a consistent distance from the camera. An improvement would be to account for this, so that the lengths of objects such as roads or very tall buildings from photographs at ground level may be measured.

Another point of improvement is accounting for rectilinear distortion, which currently affects equation accuracy at the centre and edges of a photograph.

Finally, with some further work, an equation may be derived to determine the velocities of blurry objects in a single photograph, or of the same object between two different photographs with the same field of view.

References

[1] Moxfyre, 2009. *Lens angle of View* [diagram]. Available at: http://en.wikipedia.org/wiki/File:Lens_angle_of_view.svg [Accessed 02/19/14]; modified
[2] Lens Geometries, 2010. *Photoropter*. [online] http://photoropter.berlios.de/phtrdoc/techback_geom.html [Accessed 02/19/14]
[3] Lyons, M., 2009. *Rectilinear Projection* [photograph]. Available at: http://www.tawbaware.com/projections_fed_trans_merc_rect_150.jpg [Accessed 02/19/14]
[4] Canon EOS 500D (EOS Rebel T1i / EOS Kiss X3), 2009. *dpReview*. [online] Available at: http://www.dpreview.com/products/canon/slrs/canon_eos500d [Accessed 02/19/14]
[5] Birds.kz, 2012. *Bean Goose*. [online] Available at: http://www.birds.kz/species.php?species=39&l=en [Accessed 02/19/14]
[6] Huang, C., 2014. I don't like Geese [photograph]. Unavailable online.

The Viability of Throwing Giant Tortoises onto Mines

Yannic Chen and David McDonagh
The Centre for Interdisciplinary Science, University of Leicester
17/03/14

Abstract
Due to Mario's comparatively small stature in relation to Bowser, the hero of The Mushroom Kingdom must come up with novel ways to defeat his powerful nemesis to restore order to the land. One particular method in Super Mario 64 involves throwing Bowser onto explosive mines via grabbing his tail and spinning to generate angular momentum. The viability of this technique is investigated and found to be well beyond the limits achieved by Olympic athletes.

Introduction
For decades now the popular Nintendo franchise Super Mario has been based upon the unassuming Italian plumber Mario fighting to rescue Princess Peach from The King of the Koopas and terroriser of The Mushroom Kingdom, Bowser. Although varying significantly in size throughout the series, in Super Mario 64, Bowser appears approximately three times the height of Mario, where the tortoise-like creature is grabbed by the tail and spun until enough angular velocity is generated to throw Bowser onto conveniently placed, explosive mines. Through making some basic assumptions and modelling this technique akin to the Olympic hammer throw, the viability of a human being achieving this with such a creature can be investigated.

Generating Angular Velocity
In treating Bowser as a sphere with his tail acting as a rope, the distance over which Bowser can be thrown can be represented as a projectile motion problem, where the x and y components of the displacement can be written as

$$x = v_0 \cos(\theta) t$$

$$y = v_0 \sin(\theta) - \frac{1}{2} g t^2,$$

where v_0 is the initial velocity upon release, θ is the angle upon realise, g is the gravitational force acting downward and t is time.

Rearranging the first equation for time and substituting into the second gives

$$y = \sin(\theta) \frac{x}{\cos(\theta)} - \frac{1}{2} g \frac{x}{v_0 \cos(\theta)}.$$

In taking the initial and final y values to be zero, corresponding to the height of Mario's chest and the height of the mine, this can be rearranged for velocity to give

$$\tan(\theta) x = \frac{1}{2} \frac{g x^2}{v_0^2 \cos^2 \theta}$$

$$v_0 = \sqrt{\frac{g x}{2 \tan(\theta) \cos^2(\theta)}}.$$

From a still image of the boss battle [1], the size of the arena is taken to be 30m, giving an x value of 15m if Mario is standing in the centre. Although the gravitational force of The Mushroom Kingdom is not known, this is taken to be the same as on Earth. Finally, assuming the ideal launch angle of 45° and air resistance to be negligible in the arena, this gives a required velocity of

$$v_0 = \sqrt{\frac{9.81 ms^{-2} \times 15m}{2 \tan(45) \times \cos^2(45)}}$$

$$v_0 = 12.13 ms^{-1}$$

To achieve this Mario rotates Bowser in a circle to gain angular momentum. The speed at which Mario has to rotate can be calculated by considering the angular velocity

$$v = \omega r,$$

where v is the angular velocity, ω is the angular frequency and r is the radius between Mario's centre of gravity and Bowser's. The radius r is taken to be approximately the sum of Mario's arm length and half of Bowser's height when fully extended, taken to be 0.75m and 2.25m respectively, estimated from a still image [1]. Substituting this with the calculated velocity gives

$$12.13 m\, s^{-1} = \omega(3m),$$

hence

$$\omega = 4.04\, s^{-1}.$$

The angular frequency can then be converted to revolutions per second, meaning the rate Mario will need to turn equates to:

$$2\pi f = 4.04 s^{-1}$$

$$f = 0.64\, s^{-1}.$$

This means that Mario needs to rotate roughly once every 1.38 seconds to gain enough velocity to throw Bowser 15m.

Centripetal Force
Throughout this motion, centrifugal force experienced by Mario can be calculated from the centripetal acceleration, given by

$$a = \frac{v^2}{r} = 49.05 ms^{-2}.$$

Due to the ambiguous evolutionary origin of Bowser, his mass is estimated from scaling up the mass of a giant tortoise, which based on three giant tortoises to give a similar size gives a mass of 645kg [2]. From Newton's second law, this results in a centrifugal force of

$$F = ma$$

$$F = 645 kg \times 49.05 ms^{-2} = 31634.43 N$$

pulling outward from the axis of rotation, assuming all of Bowser's mass included into the centrifugal force. Even without considering the pulling force required to accelerate Bowser to the required velocity, to counteract the centrifugal force Mario must apply an equal centripetal force to keep Bowser moving in a circle. The magnitude calculated reaches over eleven times that calculated for Olympic hammer throwers [3], which, due to Mario's comparative fitness, makes it unlikely Mario could achieve this.

Discussion and Conclusion
Despite the low frequency of rotation required to throw Bowser onto a nearby mine, such a feat would require a centripetal force beyond that of top athletes at 15m, due Bowser's mass. Decreasing this distance is not found to lower the required centripetal force to within those of hammer throwers until within the circle outlined by spinning Bowser, putting Mario himself in danger. Furthermore, the mass of Bowser means merely lifting him into a position where he could then be accelerated to the required velocity would require almost three times the current limit for what human beings can lift [4]. From this, it can be seen that it is unlikely that this method would be an efficient way of saving The Mushroom Kingdom.

References

[1] Mitjitsu, 2007, *Mario 64 beaten with 0 stars in 5:47,* Youtube, available at: http://www.youtube.com/watch?v=DTzs9bcNgMQ&t=5m22s
[2] National Geographic, 2014, *Galapagos Tortoise,* National Geography, available at: http://animals.nationalgeographic.com/animals/reptiles/galapagos-tortoise/
[3] Enoka, R.M, 2008, *Neuromechanics of Human Movement,* Sheridan Books
[4] Marshall J, 2010, *Maxed out: How much can a human lift?,* New Scientist

The Frog Prince Transformation

Yannic Chen
The Centre for Interdisciplinary Science, University of Leicester
18/03/2014

Abstract
Transformations are a common trope in fairy tales and many other media. Although they are often a core element to the story and aid in plot progression, the details about physical viabilities are often neglected. The Frog Prince (or Iron Henry) is one such example; the story features a frog turning into a prince with a significantly larger mass. While this may appear to contradict the laws of mass and energy conservation, on closer inspection, such a transformation is theoretically possible.

Introduction
The frog prince is a well-known fairy tale by the German authors Brothers Grimm. The climax of the story involves the transformation of the frog into a prince through a kiss by a spoiled princess. Since the transformation from a frog to a human comes with a large change in mass, this paper aims to investigate the possibility of such a feat with regards to the law of energy and mass conservation.

Frog to Prince
It is apparent that the mass of a frog is less than the mass of fully grown man. Considering the largest known frog species, the goliath frog (*Conraua goliath*) may grow up to 3.3 kilograms [1], and that the average mass of a male human adult is 88.7kg [2], the difference in mass is:

$$88.7 kg - 3.3 kg = 85.4 kg.$$

Assuming an instantaneous transformation and conservation of mass, the acquisition of 85.4kg has to be taken in as energy from the environment. Since mass in form of atoms and molecules cannot be readily "absorbed" by the frog and converted to organic matter, the energy equivalent is chosen instead.

Einstein's formula for mass-energy equivalence shows the relation between (rest) mass and energy [3]:

$$E = mc^2$$

Here, E is the energy, m is mass and c is the speed of light. Using this equation, the mass can be converted into an equivalent energy:

$$E = (85.4 \ kg)(2.998 \times 10^8 \ m \ s^{-1})^2 = 7.69 \times 10^{18} J$$

Putting 7.69 exajoules in perspective, the United States of America consumes 14 exajoules of electricity in 2009 [4].

Sources of the Energy
In many animated adaptions of the Frog Prince (or Iron Henry), the transformation of the frog into the prince is facilitated by a glowing body and light reaching into the body from the surroundings. The most readily available energy from the surroundings is the kinetic energy of air surrounding the frog. A simple way of finding the kinetic energy is by assuming the air behaves as an ideal gas (no in intermolecular attractive forces and perfect elastic collisions) and using the average kinetic energy (per mole of gas) formula:

$$KE_{avg} = \frac{3}{2}RT$$

Here, R is the universal gas constant and T is the temperature [5]. We will set the temperature to 20°C (293.15K), a warm German (origin of the authors) spring day. The average kinetic energy per mole will be:

$$KE_{avg} = \frac{3}{2}(8.3145\ J\ mol^{-1}K^{-1})(293.15K) = 3656\ J\ mol^{-1}$$

In order to gain the amount of energy required for a successful transformation, 2.10×10^{15} moles of gas are required. Since the molecular mass of air is $0.02897\ kg\ mol^{-1}$ [6], the required mass is therefore:

$$2.1 \times 10^{15} mol \times 0.02897\ kg\ mol^{-1} = 6.08 \times 10^{13} kg$$

Considering that the total mean mass of the atmosphere is $5.15 \times 10^{18} kg$ [7], the effect of the transformation would span an area of approximately 1.1×10^{-5} times the Earth.

Conclusion

It is theoretically possible to gain enough energy for the transformation of a frog to a prince from the kinetic energy of air alone. However, extraction of such amount of energy is practically impossible for present science, as it would require extracting all kinetic energy from the air molecule. This would result in an environment of absolute zero temperature. While solar heating would slowly provide the air with energy again, the health of the prince/frog and any other person (princess) in the area cannot be assured. Considering that story was written and set in an age more than a hundred years ago, such a feat can truly only be achieved with magic.

References

[1] San Diego Zoo. (no date). *Goliath frog*. Accessed February 14, 2014, from San Diego Zoo Animals: http://animals.sandiegozoo.org/animals/goliath-frog
[2] CDC/National Center for Health Statistics. (2012). *Bod Measurements.* Accessed February 14, 2014, from Center for Disease Control and Prevention: http://www.cdc.gov/nchs/data/series/sr_11/sr11_252.pdf
[3] Nave, R. (no date). *Relativistic Energy*. Accessed February 14, 2014, from Hyperphysics: http://hyperphysics.phy-astr.gsu.edu/hbase/relativ/releng.html
[4] Central Intelligence Agency. (2014). *The World Factbook*. Accessed February 14, 2014, from https://www.cia.gov/library/publications/the-world-factbook/geos/us.html.
[5] Nave, R. (no date). *Ideal Gas Law*. Accessed February 20, 2014, from Hyperphysics: http://hyperphysics.phy-astr.gsu.edu/hbase/kinetic/idegas.html#c1
[6] The Engineering Toolbox. (no date). *Molecular Mass of Air*. Accessed February 20, 2014, from The Engineering Toolbox: http://www.engineeringtoolbox.com/molecular-mass-air-d_679.html
[7] Trenberth, K. E., & Smith, L. (2003). *The Mass of the Atmosphere: a Constraint on Global Analysis*. Retrieved February 20, 2014, from UCAR: http://www.cgd.ucar.edu/cas/abstracts/files/kevin2003_6.html

A Scandal in Belgravia ... for whom?

Stephanie Taylor
The Centre for Interdisciplinary Science, University of Leicester
20/03/2014

Abstract
This paper investigates the observations and deductions made by Sherlock Holmes in the BBC television show *Sherlock*. Sherlock makes many incredible observations to solve his cases, including spotting dog hairs on a pant leg across the room. This paper uses the angular resolution to determine if Sherlock's observations are scientifically plausible.

Introduction
Sherlock Holmes is a fictional detective, created by Sir Arthur Conan Doyle in 1887. Recently, the BBC has recreated the Sherlock Holmes novels as a television show, set in the present. The show, titled *Sherlock* has become quite popular, and it chronicles the adventures of Sherlock Holmes and Dr. John Watson. Throughout the show, Sherlock makes a number of astounding deductions to help solve London's most difficult mysteries. This leads to the question: can Sherlock actually make these deductions? Or is the BBC bending human limitations to make a good television show?

This paper looks into one particular deduction Sherlock made in the first episode of the second season, titled *A Scandal in Belgravia*. In the episode, members of the Royalty Protection Command arrive at 221B Baker St. to transport Sherlock to Buckingham Palace for a client consultation. However, the men do not tell Sherlock who they are; he deduces it by making observations. One specific observation was the presence of dog hairs on the pant leg of one of the men. Could Sherlock actually see those dog hairs?

Angular Resolution
To determine if Sherlock Holmes could indeed observe the dog hairs, the angular resolution of the human eye and the angular diameter of dog hairs at that distance needs to be determined. The latter may be accomplished using the small angle approximation, which is a variation of the equation used by astronomers to determine the angular diameter of celestial objects [1]:

$$\theta = 206265 \left(\frac{d}{D}\right)$$

In this case, d is the width of a dog hair and D is the distance between Sherlock's eyes and the dog hairs. A study conducted in 2009 determined that the average width of a dog hair is 25μm (2.5x10^{-5}m) [2]. D is determined using Pythagoras' Theorem, the vertical distance between Sherlock's eyes and the dog hairs (a) and the horizontal distance between the two men (b) as shown in Figure 1.

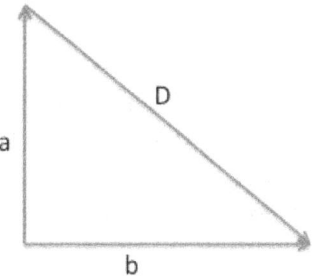

Figure 1: A representation of the measurements used to determine D, where a is the vertical distance between Sherlock's eyes and the dog hairs, and b is the horizontal distance between the two men.

Using a still from the episode, b is estimated to be 5ft or ~1.5m. Since Sherlock is sitting down when he makes the observations, to find a, his sitting height must be found. Sherlock's height when standing is 1.83 m [3]. The height in question is the height of Benedict Cumberbatch, the actor who plays Sherlock. A recent study found the ratio of sitting to standing height in adult males is 0.52, which makes Sherlock to be about 0.957m tall when sitting [4]. The last factors to account for are the height of Sherlock's eyes (~10cm lower) and the height of the dog hairs (~15cm off the ground) both of which were estimated from the still of the episode. This results in a measurement of 0.61m for *a*. Using Pythagoras' Theorem, the value of *D* is around 2.6m.

With values for *d* and *D*, the angular resolution of the dog hairs can be determined using the small angle formula:

$$\theta = 206265 \left(\frac{d}{D}\right)$$

$$\theta = 206265 \left(\frac{2.5 * 10^{-5} m}{2.6 m}\right)$$

$$\theta = 2''$$

Therefore the angular resolution of the dog hairs is close to 2 arcseconds. The angular resolution of the human eye, or the smallest value of θ that can be seen by the human eye, is 50 arc seconds [5]. Since the angular resolution of dog hairs well below angular resolution of the human eye, it was not possible for Sherlock to observe the dog hairs.

Conclusion

While basic principles of angular resolution have determined that it was not possible for Sherlock to deduce the man spent a lot of time with dogs, there are other factors involved. In this investigation, the width of the dog hairs was used as opposed to the length. In a further study, the angular resolution of the dog hair should be calculated using their length. This will account for the possibility that the length of the dog hairs could have allowed Sherlock to see them.

In addition to angular resolution of the dog hairs, there are many other factors that could have contributed to Sherlock's observation. Further investigation into this deduction should look at the effects of colour contrast and quantity of hairs present on visual resolution. It is possible that these factors could have allowed Sherlock to see the dog hairs.

It appears that the real scandal in Belgravia was caused by Sherlock Holmes.

References

[1] Anon, 2014. *Small Angle Formula*. [online] Imaging the Universe. Available at: http://astro.physics.uiowa.edu/ITU/glossary/small-angle-formula/ [Accessed 7 Feb. 2014].

[2] Kshirsagar, S.V., Singh, B. and Fulari, S.P., *Comparative Study of Human and Animal Hair in Relation with Diameter and Medullary Index*. Indian Journal of Forensic Medicine and Pathology, **2**(3), pp.105–8 (2009).

[3] Benedict Cumberbatch Height – How tall. 2014. *Benedict Cumberbatch Height – How tall*. [ONLINE] Available at: http://www.celebheigts.com/s/Benedict-Cumberbatch-2886.html. [Accessed 8 Feb. 2014].

[4] De Arriba Munoz, A., Dominguez Cajal, M., Rueda Caballero, C., Labarta Aizpun, J.I., Mayayo Dehesa, E. and Ferrandez Longas, A., *Sitting/standing height ratio in Spanish children from birth to adulthood*. Archivos Argentinos De Pediatria, **111**(4), pp.309–314 (2013).

[5] Huard, T., n.d. *Angular Resolution*. Available at: http://www.astro.umd.edu/~thuard/astr288c/lecture6-notes.pdf [Accessed 8 Feb. 2014].

Golden Fleece: A Heavy Task

Sean Gilmore
The Centre for Interdisciplinary Science, University of Leicester
24/03/2013

Abstract
Of the many infeasible creatures and relics in ancient Greek mythos, the Golden Fleece from Jason and the Argonauts has drawn much attention from historians as to what it represented in terms of politics, technology and religion. However, we will instead explore the scientific basis to the possibilities of the existence of a gold fleeced ram. This article specifically addresses the physical and biophysical aspects to this multidisciplinary problem.

Introduction
The ancient Greek tale of Jason and the Argonauts is one of the most famous and foundational in classical Greek history. In one of its oldest and most complete accounts, the 3rd century BC *Argonautica* by Apollonius Rhodius, the hero Jason quests across uncharted seas and overcomes impossible obstacles in order to obtain a legendary relic, with promise of his rightful crown. This relic, found in the land of Colchis (the eastern Caucasus), is none but the fleece of a ram with hair of gold. Although a golden fleeced ram borders on the ordinary among the many fanciful creatures and relics in the tales of ancient Greece, it is the fleece's representation of power that is at the heart of Jason's quest. However, if such a creature were to exist, it would raise a number of issues regarding physical, geochemical, biochemical, and evolutionary feasibility.

Fleece Weight
The fleece of a ram, like the fur and hair of other mammals, is made primarily of keratin [1]. In comparison, the Golden Fleece will be modelled to have a primary composition of gold. In order to calculate the weight of the fleece, the density ratio of the constituent keratin and gold is assumed to be equal to that of the woollen and aurous (made of gold) wool. In essence, this is an assumption that there is the same volume of gold per volume of aurous wool as there is keratin per volume of ordinary wool. It can then be said that this ratio would be equal to the ratio of the fleece weights, as shown in the equation below, since the volume components cancel.

$$\frac{\rho_{ker}}{\rho_{Au}} = \frac{m_{ker}}{m_{Au}}$$

Here, ρ_{ker} is the density of keratin, roughly 1280 to 1340 kg m^{-3} [1], and ρ_{Au} is the density of gold, at 19320 kg m^{-3} [2]. m_{ker} and m_{Au} are the respective weights of a woolen and aurous fleece.

In the days of 3rd and 4th century BC Eastern Europe, sheep would certainly have looked rather different to today's industry-driven, highly modified stock. However, we cannot assume that the sheep used in civilised ancient Greek agriculture would be similar to purely feral or wild sheep, such as the Armenian mouflon [3]. A group of nomadic peoples, the Sarakatsan, originating from ancient Greek tribes, preserved the culture of ancient Greek shepherding in areas of the Baltics until the last century [4]. Ancient Greek agricultural rams are likely to have resembled those of the Sarakatsaniko sheep, which average 66 kg and have a mean wool weight of 3.5 kg.

Taking this wool weight as m_{ker} gives the Golden Fleece a weight of 50 to 53 kg. At the weight of a small person, the depictions of the Golden Fleece, being easily adorned on the shoulders of Jason and his companions, are surely inaccurate.

Maximum Carry Weight
Regarding the poor animal itself, carrying such a heavy fleece during its whole life before maturation and eventual slaughter may cause issues regarding the biophysical limits to ungulate carry weight. Research has been carried out on pack-animals such as horses, mules and goats, while not on other herd-animals that are not typically used for this purpose, such as sheep. Nevertheless, this information may be used to get a reasonable idea as to the weight

the ram could withstand. Equine veterinary scientists have shown that a horse carrying more than 20% of its weight is likely to cause damage to muscle and joints [5]. Some sources, however, quote the remarkable strength of mules carrying comfortably over 30% of their body weight [6]. The smaller pack-goat, more similar to our ram, is said to also be able to carry a third of its weight [7]. The Golden Fleece is calculated to weigh around 80% of the mass of the average Sarakatsaniko ram [4]. It therefore needs the strength of a ram over double its size in order to carry its own fleece. A larger sheep will have a relatedly increased fleece weight, and an even heavier task.

Conclusion

Research separately carried out by this group suggests that the golden wool would be up to 1000 times more thermally conductive than ordinary wool. The sheep would require a much thicker fleece to survive in the Caucasus, where temperatures in lowland regions remain negative throughout winter [8]. This, along with the proportionality issue mentioned above, suggests that the 50 kg fleece is likely to be a severe underestimate. This indicates the extremely low likelihood that such a creature could survive to maturity, as being in perpetual pain and exhaustion would make it unable to flock and evade predation successfully.

References

[1] R. D. B. Fraser and T. P. MacRae, "Molecular organization in keratins. II. Densities of native keratins," Textile Research Journal, vol. 27, no. 384, 1957.

[2] The Engineering ToolBox, "Metals and Alloys - Densities," [Online]. Available: http://www.engineeringtoolbox.com/metal-alloys-densities-d_50.html. [Accessed 23 February 2014].

[3] Oklahoma State University Board of Regents, "Sheep: (Ovis aries)," 2008. [Online]. Available: http://www.ansi.okstate.edu/breeds/sheep/. [Accessed 24 February 2014].

[4] D. Dervisis, "The Sarakatsankio Sheep," [Online]. Available: http://www.agrobiodiversity.net/greece/pdf/Sarakatsaniko_Dervisis.pdf. [Accessed 24 February 2014].

[5] D. M. Powell, K. Bennett-Wimbush, A. Peeples and M. Duthie, "Evaluation of Indicators of Weight-Carrying Ability of Light Riding Horses," Journal of Equine Veterinary Science, vol. 28, no. 1, pp. 28-33, 2008.

[6] W. Carlton, Interviewee, More with Wayne Carlton on Elk Hunting. [Interview]. 2012.

[7] C. Zimmerman, "Frequently Asked Questions," High Uinta Pack Goats, [Online]. Available: http://www.highuintapackgoats.com/FAQ_page.html. [Accessed 8 March 2014].

[8] Geographic Bureau, "Climate in the Caucasus," Geographic Bureau, 2014. [Online]. Available: http://www.elbrus.su/info/climate_in_the_caucasus. [Accessed 21 March 2014].

Evaluating *The Core*: The prospect of geodes

Kira Moor and Rebekah Ingram
Honours Integrated Science, McMaster University
24/03/2014

Abstract
This paper analyses the legitimacy of some of the science behind the movie *The Core*. The feasibility of the formation of large amethyst geodes in the Earth's mantle as seen in *The Core* is investigated by calculating the pressure in the mantle and analyzing the effect of mantle temperature on amethyst crystallization. The analysis shows it to be extremely unlikely that geodes exist in the lower mantle.

Introduction
Released in 2003, *The Core* is a science fiction film in which the convective currents in the Earth's liquid outer core stop. A group of scientists must travel through the Earth's crust and mantle towards the centre of the Earth and attempt to restart convection in the outer core. During their journey, the scientists are amazed when their ship becomes damaged as they travel through a large amethyst geode in the mantle.

Amethyst Geode Formation
A geode is a geological formation which is typically spherical, with a plain outer crust of igneous or sedimentary rock, followed by an inner cavity space partially or completely filled with crystallized minerals [1].

Amethyst bearing geodes appear in many regions around the world. The highly studied amethyst geodes of Rio Grande do Sul, Brazil can be used as an analog for their formation. The Rio Grande do Sul geodes were likely formed in a two-step process. The early stage 'protogeode' cavity was created as gases were released by cooling lava, and then trapped within the hot, molten rock. The later stage cavity infilling would have then occurred through the flow of mineral rich fluids in the cavity [2]. These fluids were expected to have been low temperature, gas-poor, aqueous solutions of very low salt content [2].

Conditions necessary for amethyst geode formation therefore include the presence of a cavity and low temperature, mineral rich fluid, in addition to the conditions necessary for amethyst crystallization.

Mantle Cavity
Within the mantle, the force of the overlying rocks would create a high amount of stress which would act against the formation of a cavity. The magnitude of stress present at the depth in the mantle where the amethyst geode was found is dependent on the density and thickness of the overlying rock, as well as the gravitational acceleration acting upon this rock. This stress can be approximated using the following equation [3]:

$$\sigma = \rho g z$$

where $\rho = density$
$g = gravitational\ accelleration$
$z = depth\ below\ Earth's\ surface$

More simply, predetermined lithostatic stress gradients use average material density multiplied by the gravitational acceleration to replace these terms in the above equation [3]. For the Earth's crust, the lithostatic stress gradient is 26.5 MPa/km, while the gradient for the mantle is 35 MPa/km [3]. The geodes in the movie are found 700 miles, or greater than 1000 km, below the surface of the Earth. The total lithostatic stress experienced at this depth would be:

$$\sigma_{crust} = 40\ km \times lithostatic\ stress\ gradient$$
$$\sigma_{mantle} = 1078\ km \times lithostatic\ stress\ gradient$$
$$\sigma_{total} = 40\ km \times 26.5\ MPa/km + 1087\ km \times 35\ MPa/km$$
$$\sigma_{total} = 39105\ MPa$$
$$\sigma \approx 39\ GPa$$

The closest analogue we have at the Earth's surface for mantle material is olivine rich basalt. Under pure stress conditions caused by the weight of overlying rock, less than 300 Pa would be needed to overcome the strength of the crystal network in basalt and induce melting [4]. Thus, under the extreme stress conditions of approximately 39 GPa experienced at a depth of over 1000 km, minerals in the mantle would undergo ductile deformation and would flow rather than form the solid crust of a hollow sphere. This makes it highly unlikely that a cavity in which amethyst crystals could form would be located deep within the Earth's mantle.

Amethyst Crystals
The formation of amethyst crystals in the mantle may seem acceptable, maybe even probable, but this is not the case. Amethyst is the violet variety of alpha-quartz with its slight purple colouring attributed to trace amounts of iron in the crystal composition [2]. From alpine clefts to miaroles in granite rock, amethyst crystals form in various geologic environments [2]. It is important to understand the characteristics of each setting to determine whether the growth of these crystals is supported.

Different temperatures have been attributed to the growth of amethyst crystals. In the amethyst mines of Thunder Bay, Ontario the crystals were determined to form in 40-90 degrees Celsius temperatures [2]. Brazilian amethyst formation temperatures ranged from 50-100 degrees; with a general consensus they did not exceed 100 degrees Celsius. In Scotland, amethyst formation temperatures are expected to have been less than 150 degrees Celsius. These temperatures are expected to have been between 150-250 degrees in Hungary and have reached 280-400 degrees in South Korea [2].

Individual quartz grains form a tetrahedron of four shared oxygen atoms [5]. Due to the shared nature of the atoms, increasing temperatures causes this framework silicate to break down into simpler silicate forms. Using Bowen's Reaction Series, it can be determined that the ideal temperatures for quartz to form are below 750 degrees Celsius (Figure 1) [5]. Above these temperatures more mafic minerals are expected to form such as biotite, amphibole and pyroxene. In the movie, the crew finds the amethyst geode in the lower mantle. While the temperature of the lower mantle is unknown, plate tectonic models have estimated mantle temperatures as high as 3726 degrees Celsius [6]. These drastically high temperatures illustrate the unlikelihood of quartz and therefore amethyst crystals forming so deep within the Earth.

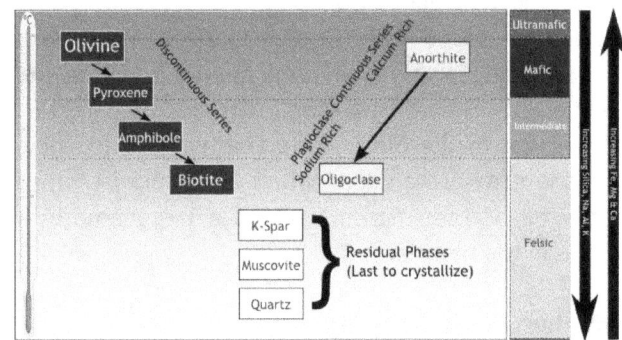

Figure 1: Bowen's Reaction Series showing the lower temperatures at which quartz crystallizes [7].

Conclusion
While a geode of amethyst crystals offers visual appeal and interest to the film, the chance of finding this geologic formation deep within the mantle is highly unlikely. The stress at this depth would be too great for mantle rocks to form a solid shell for a cavity. Additionally, the formation of amethyst crystals deep in the mantle would be impossible due to very high temperatures.

References

[1] Acton, Q.A., 2013. *Silicon Compounds - Advances in Research and Application.* Atlanta: ScholarlyEditions.
[2] Gilg, H.A., Morteani, G., Kostitsyn, Y., Preinfalk, C., Gatter, I., and Strieder, A.J., *Genesis of amethyst geodes in basaltic rocks of the Serra Geral Formation (Ametista do Sul, Rio Grande do Sul, Brazil): a fluid inclusion, REE, oxygen, carbon and Sr isotope study on basalt, quartz and calcite,* Mineralium Deposita, **38**, pp. 1009-1025 (2003).
[3] Fossen, H., 2010. *Structural Geology.* New York: Cambridge University Press.
[4] Hoover, S.R., Cashman, K.V., and Manga, M., *The yield strength of subliquidus basalts - experimental results,* Journal of Volcanology and Geothermal Research, **107**, pp. 1-18 (2001).
[5] Stanley, S.M., 2008. *Earth System History.* Third edition. New York: W.H. Freeman.

[6] Hirose, K., Fei, Y., Ma, Y., and Mao, H.K., *The fate of subducted basaltic crust in the Earth's lower mantle*, Nature, **397**, pp. 53-56 (1999).
[7] Colovine, 2011. *Bowen's Reaction Series*, http://commons.wikimedia.org/wiki/File:Bowen%27s_Reaction_Series.png [Accessed 12 March 2015].

Journal of Interdisciplinary Science Topics

The Winter Olympics on Enceladus

David McDonagh
The Centre for Interdisciplinary Science, University of Leicester
24/03/2014

Abstract
The viability of Enceladus as a future location for The Winter Olympic Games is considered, using the large hill ski jumping event of the Sochi 2014 slope as a case study. Different conditions are found to significantly increase descent and air time, while the lack of a significant atmosphere could lead to interesting developments in the sport.

Introduction
Despite the successes of the Winter Olympic Games over the last century, increasing costs and political tensions highlight the need for neutral ground if the event is to continue into the future. Looking out into the solar system, Saturn's icy moon Enceladus is an obvious choice. Blankets of snow up to 100 metres thick are estimated in some regions [1], made of the finest particles, making it ideal for winter Olympic events. To assess the viability of such a proposal, the ski jumping large hill event is investigated as an example of how Olympic athletes would need to adapt to compete on this distant moon.

Skiing on Enceladus
Aside from the need to wear space suits, differences in gravity and atmospheric density are likely to be the key factors affecting skiing events. The surface gravity of Enceladus is 0.113 ms^{-2}, approximately 1.15% that of Earth [2], while the atmospheric density is currently approximated to be around 2.99×10^{-12} $kg\ m^{-3}$, based on Cassini data and computational simulations [3]. For simplicity, an assumption is made here of an atmosphere composed entirely of water vapour (the measured value is 91% [3]). As this value is approximately a trillion times smaller than the atmospheric density on Earth, air resistance is deemed negligible. To give a good indication of the differences on Enceladus compared to on Earth, the dimensions of the Olympic hill of the recent large hill ski jumping event in the 2014 Sochi Olympics are used [4].

Assuming the skier starts at the maximum in-run length (the top of the slope), upon release the skier will begin to accelerate due to gravity, slowed by friction between the skis and the snow (figure 1). The total force (F_{Total}) will hence be

$$F_{total} = F_{parallel} - F_{friction}.$$

Figure 1: The forces acting on the skier when descending the slope. Image credit: Wikipedia (modified).

If the mass of the skier is taken to be 70kg and the slope angle is taken to be 35° [4], the force parallel to the slope will be

$$F_{parallel} = ma\sin(\theta)$$
$$F_{parallel} = 70kg \times 0.113ms^{-2} \times \sin(35°)$$
$$F_{parallel} = 4.537N.$$

The friction force is given by

$$F_{friction} = F_N \mu_k,$$

where μ_k is the coefficient of kinetic friction, which for waxed skis on snow is 0.05 [5] (an approximation is made here as this is the value at 273K with Earth's gravity, while Enceladus is roughly 72K [2], hence this is likely a lower limit).

$$F_{friction} = ma\cos(\theta) \times 0.05$$

$$F_{friction} = 70kg \times 0.113ms^{-2} \times \cos(35°) \times 0.05$$
$$F_{friction} = 0.324N,$$

giving a net force of

$$F_{total} = 4.213N.$$

The acceleration of the skier is then

$$a = \frac{F_{total}}{m}$$
$$a = \frac{4.213N}{70kg} = 0.060ms^{-2}.$$

The velocity at the bottom of the slope can be found using

$$v_f^2 = v_i^2 + 2as,$$

where s is the length of the slope, v_i is the initial velocity and v_f is the final velocity. Using 99.3m as the length of the slope [4], and substituting this and the value for acceleration gives

$$v_f^2 = 0 + 2(0.060ms^{-2} \times 99.3m)$$
$$v_f = 3.452 \ ms^{-1}.$$

The skier takes off at an angle of 11° and a height of 3.14m [4], after which the only force acting on them is gravity. Change in height over time can be calculated using

$$y = v_0^y t + \frac{at^2}{2}, + y_0$$

where y_0 is the initial height, and v_0^y is the initial vertical velocity, given by

$$v_0^y = 3.452ms^{-1} \times \sin(11°) = 0.659ms^{-1}.$$

Plotting values of the skier's height against time gives the graph shown in *figure 2*, where the skier lands after approximately 9 seconds. The horizontal displacement will be the horizontal velocity multiplied by time, resulting in a horizontal displacement of

$$x = 3.452ms^{-1} \times \cos(11°) \times 9s$$
$$x = 30.5m.$$

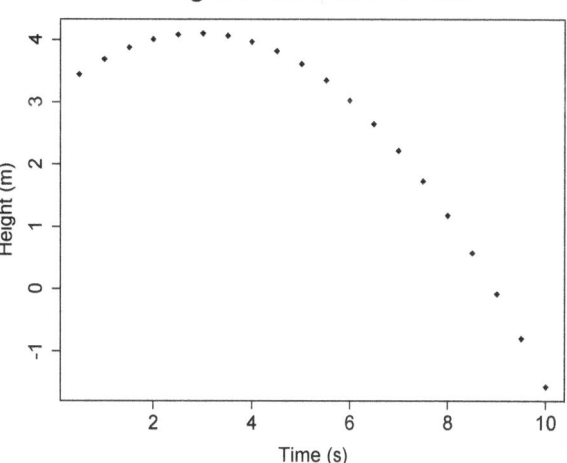

Figure 2: The height of the skier over time once in the air.

Discussion and Conclusion

In comparing the calculated results to the 2014 Sochi Olympics [6, 7], an average distance reached would only be a fifth of that seen on Earth. Furthermore, dividing the take off velocity by the acceleration shows around 30 seconds would be needed to reach the bottom of the slope, compared to around 5 seconds as seen at Sochi [6], while the time spent in the air nearly doubles on Enceladus. Such changes would likely result in a much calmer, safer sport. Interestingly, the lack of a significant atmosphere would make friction the only component dependent on the athlete, which would depend on the athlete's mass. This would likely shift the attention away from a focus on reducing drag to that of reducing friction, perhaps encompassing techniques used in curling in some way, giving just one example of how The Winter Olympic Games can have a bright future on the brightest moon in the solar system.

References

[1] Schenk, P., Schmidt, J., & White, O., (2011), EPSC Abstr, 6, EPSC-DPS2011-1358
[2] NASA, (n.d), Enceladus: Facts & Figures, NASA, available at:
http://solarsystem.nasa.gov/planets/profile.cfm?Object=Sat_Enceladus&Display=Facts

[3] Tenishev, V., Combi, M.R., Teolis, B.D., & Waite, J.H., (2010), An approach to numerical simulation of the gas distribution in the atmosphere of Enceladus, Journal of Geophysical Research, 115

[4] Ski jumping Archive 2002-2014, (2014), RisSki Gorki Jumping Center, Ski Jumping Hill Archive, available at: http://www.skisprungschanzen.com/EN/Ski+Jumps/RUS-Russia/Esto-Sadok/0415-RusSki+Gorki+Jumping+Center/

[5] Tipler P.A., & Mosca G., (2008), Physics for Scientists and Engineers, W.H Freeman and Company, p132

[6] Olympics Youtube Channel, (2014), Ski Jumping – Men's Large Hill Final, Olympic.org, available at: http://www.youtube.com/watch?v=x9IuBMEcIuc

[7] Sochi 2014, (2014), Men's Large Hill Individual Final Round, Sochi 2014, available at: http://www.sochi2014.com/en/ski-jumping-men-s-large-hill-ind-final-round

How many lies could Pinocchio tell before it became lethal?

Steffan Llewellyn
The Centre for Interdisciplinary science, University of Leicester
25/03/2014

Abstract:
This paper investigates how many lies Pinocchio could continuously tell before it would become fatal, treating the head and neck forces as a basic lever system with the exponential growth of the nose. This paper concludes that Pinocchio could only sustain 13 lies in a row before the maximum upward force his neck could exert cannot sustain his head and nose. The head's overall centre of mass shifts over 85 metres after 13 lies, and the overall length of the nose is 208 metres.

Pinocchio's Nose

Pinocchio is the fable of a wooden puppet, carved by Geppetto, who dreams of becoming a real boy [1]. Pinocchio was portrayed as a character prone to lying, which is manifested physically through the ability to grow his nose when he tells a lie. One issue of growing his nose would be the shift of Pinocchio's centre of mass within his head, causing strain on his neck, which helps stabilise his head's position with upwards force. If this continued, then his neck could not support his head, potentially decapitating the puppet. Outlined here is the minimum lie count Pinocchio could continuously expel. Where Pinocchio manages to form new is not addressed in this paper.

Maximum Force Pinocchio's Neck Can Exert

The assumption is simplified by allowing the force exerted upwards through the neck to be positioned at the back of the head. The head is treated as a sphere, and the nose as a cylinder, as shown in Figure 1.

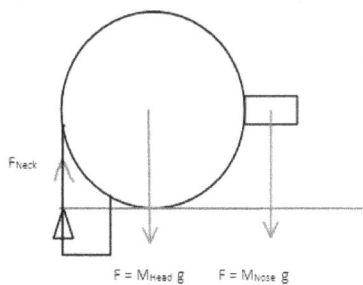

Figure 1: Illustrates the lever system of Pinocchio's head and neck, with opposite forces.

The type of wood Pinocchio is carved from is disputed, but for this paper, it is concluded that Pinocchio is made from Oak, with a density of $\approx 750\ kgm^{-3}$. Pinocchio's neck will brake if its compression strength threshold is overcome by the weight of his head. The compression strength of oak is $1150 Psi \approx 7.9 \times 10^6 N$ [2], and the circumference of the average human neck is 0.4m [3]. The maximum force Pinocchio's neck can sustain is:

$$Circumference = 2\pi r$$
$$r = \frac{0.4}{2\pi} = 0.0637m$$
$$Pressure = \frac{Force}{Area}$$
$$Force = Pressure \times Area$$
$$= (7.9 \times 10^6) \times \pi(0.0637^2)$$
$$Force \approx 1 \times 10^5 N$$

Centre of Mass, and Force Exerted

Neck muscles are required to balance the weight exerted by the skull. Usually, the weight of the nose can be considered negligible. In Pinocchio's case, as the nose increases, it will have a significant impact on the centre of mass and weight of his head. The mass of the head is unchanged:

$$Mass\ of\ Head = Density \times Volume$$
$$= 750 kgm^{-3} \times \frac{4}{3}\pi(0.11^3).$$
$$Mass\ of\ Head = 4.18 kg.$$

The nose initially can be considered negligible. However, it becomes more significant as it increases in size. For this model, the nose has an initial length of 1 inch (2.54cm), a diameter of 2cm, and its centre of mass positioned in the middle:

$$Mass\ of\ nose = Density \times Volume$$
$$= 750 kgm^{-3} \times \pi(0.01^2)(0.0254)$$
$$Mass\ of\ nose = 0.006 kg$$

Such a small mass does not affect the Centre of Mass, nor does it apply great force on the neck.

$$Centre\ of\ Mass = \frac{M_H x_H + M_N x_n}{M_H + M_N} =$$
$$\frac{[(4.18 \times 0.11) + (0.006 \times 0.1227)]}{4.18 + 0.006}$$
$$CoM = 0.11m$$

Since this is a lever system, the weight applied on the neck also depends on the distance from which that force is being applied. The initial force Pinocchio's head exerts is:

$$Total\ Force\ Applied = [(M_H g \times l_h)] + [(M_N g \times l_n)].$$
$$Total\ Force\ Applied = [(4.18 \times 9.81 \times 0.11)] + [(0.006 \times 9.81 \times 0.1227)]$$
$$Total\ force\ Applied = 4.51N$$

This force is miniscule in comparison to the strength of Pinocchio's oak neck, thus, there is no great pressure applied on Pinocchio.

Growth of the nose

During Disney's Pinocchio, the puppet's lies cause extreme growth of his nose [4]. Thus, it is not absurd to model one lie causing an increase in the length of the nose by a factor of two (i.e. the nose doubles in length for every lie). Assuming the nose also remains the same density, as mass is in proportion with the volume, the length of the nose will eventually exert significant force to the head-neck lever system. The nose is determined unbreakable, as it built upon the foundation of lies. *Figure 2* demonstrates the force exerted by the head on the neck. Once Pinocchio's nose grows to the point at which it exceeds just over 140 metres, the force exerted downwards would cause the supporting neck to snap. Due to the exponential nature of his nose growth, Pinocchio cannot tell 13 consecutive lies, as, at this point, his nose would reach a length of 208m, and his centre of mass would have shifted by roughly 85m. *Table 1* outlined in *Supplementary material* demonstrates calculations made for this approximation.

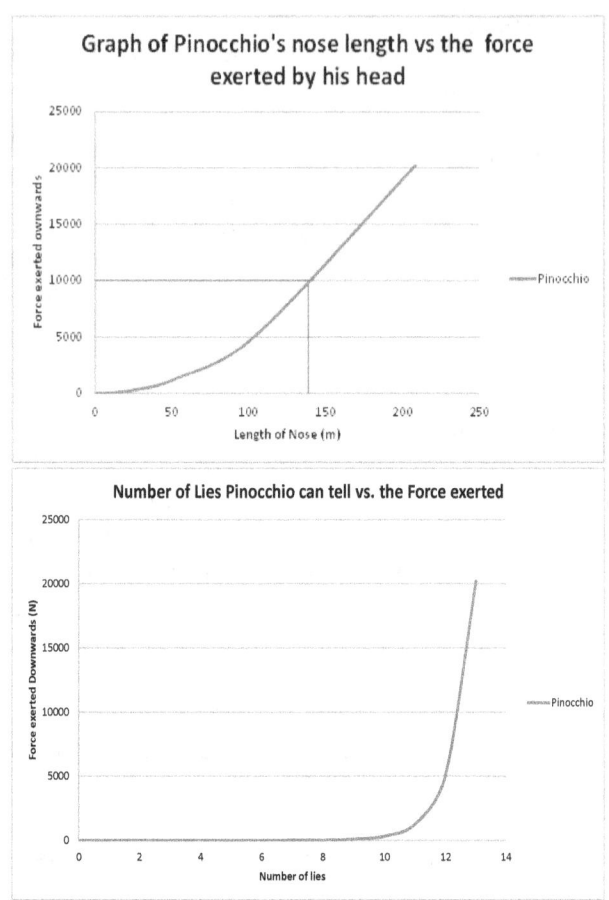

Figure 2a: Graph demonstrating how the growth of Pinocchio's nose would alter the mass to such an extent, it would overcome the forces of his neck.
Figure 2b: Graph demonstrating the number of lies Pinocchio could tell vs the total downward force exerted. It demonstrated an exponential relationship.

Conclusion

Pinocchio's anatomy has extraordinary properties. It is outstanding that his nose seemingly deposits mass from nowhere. Nevertheless, this unique ability can be of great concern for the puppet, and lengthy, extensive lies are advised against, for the health and well-being of Pinocchio.

References

[1] DisneyWiki, "Pinocchio," 2013. [Online]. Available: http://disney.wikia.com/wiki/Pinocchio. [Accessed March 2014].
[2] Engineering Toolbox, "Wood Beams Strength," 2014. [Online]. Available: http://www.engineeringtoolbox.com/wood-beams-strength-d_1480.html. [Accessed March 2014].

[3] Anonymous, "What is the average circumference of the human neck?," December 2013. [Online]. Available: http://www.chacha.com/question/what-is-the-average-circumference-of-a-human-neck. [Accessed March 2014].

[4] W. Disney, "Pinocchio 70th Anniversary Platinum Edition - Pinocchios Lies," Disney, 2009. [Online]. Available: https://www.youtube.com/watch?v=hJ3lxzuI_sc. [Accessed March 2014].

Journal of Interdisciplinary Science Topics

BoRK or BT? An Analysis for Vayne Players in League of Legends

Chuqiao Huang
The Centre for Interdisciplinary Science, University of Leicester
25/03/2014

Abstract
Since the Blade of the Ruined King (BoRK) received significant stat decreases on League of Legends, Vayne players have increasing begun to purchase the Bloodthirster (BT) instead. This paper aims to address when buying one item over the other is optimal through finding the difference between kill times on targets with varying health and armour in a plausible mid-game scenario with each item.

Introduction
League of Legends (LoL) is a multiplayer game published by Riot Games. Players control a character known as a champion and work in teams of five to take down the structure of an opposing team known as the "Nexus" [1].

One currently popular champion is Vayne, who is played in 18.23% of all games [2]. In August of 2013, an item universally utilized by Vayne players known as the "Blade of the Ruined King" (BoRK) received stat decreases [3]. This raised the question of whether the item should be purchased anymore over the "Bloodthirster" (BT), a similar alternative.

This article aims to identify when purchasing one item over the other is preferable by modelling a plausible mid-game scenario: at 16 minutes, a fight has broken out around dragon. By creating surface functions for kill times on targets with varying armour and health for each item and subtracting the two functions, the effectiveness of each item can be determined.

Defensive Stats
We will first consider the stats that govern a target's survivability. While there are many such stats in LoL, only two are relevant to our model: hit points (HP) and armour (AR).

HP is the amount of damage a target can receive before dying.

AR reduces incoming physical damage by a multiplier (AR_{red}) [4]:

$$AR_{red} = \frac{100}{100 + AR}$$

AR is affected by attacker armour penetration (APen), which ignores a percentage of a target's AR. Taken together, physical damage on a per hit basis is multiplied by [4]:

$$AR_{red} = \frac{100}{100 + (1 - Apen)AR}$$

We will consider two types of damage for our calculations: physical damage, which is mitigated by armour, and true damage, which is not.

Offensive Stats
We will now review the stats that affect Vayne's damage per second (DPS). Although there are many factors affecting DPS, we will only consider attack damage (AD), attack speed (AS), and "Silver Bolts". AD is the physical damage inflicted per hit [5] and AS is the number of attacks per second [6]. With this in mind, DPS is:

$$DPS = AS \times AD$$

By accounting for decreases in physical damage due to armour, the DPS formula becomes:

$$DPS = AS(AR_{red} \times AD)$$

Vayne deals true damage on every third consecutive hit through an ability known as "Silver Bolts"; the damage is dependent on the number of "points" invested in the skill. In the scenario given, this value is 30 + 5% of a target's maximum health (HP_{max}) [7]. For simplicity, we will incorporate this into our model by dividing the damage dealt by three and applying it to every hit. Accounting for silver bolts, the DPS formula becomes:

$$DPS = AS(AR_{red} \times AD + 10 + \frac{HP_{max}}{60})$$

Finally, the time to kill a target in seconds (k_t) may be determined by dividing the maximum health (HP_{max}) by the DPS:

$$k_t = \frac{HP_{max}}{DPS}$$

While Vayne has additional damage increasing skills, we will not incorporate them into our model. In addition, all champions have health regeneration (which decreases effective DPS); for the sake of simplicity, we will also ignore this.

Offensive Statistical Modifiers
There are many methods by which a Vayne player may increase their AD and AS. These may be divided into pregame and in-game modifiers.

Pregame modifiers consist of runes and masteries, which give a persistent bonus to AD and AS. To determine the effects of runes and masteries on our model, we will examine a Vayne game played by a professional LoL player in Feburary 2014. The net effect of his runes and masteries were: +14.76 AD, +0.55 AD/level, +5% bonus AD, +0.0329 AS, +6% Apen, +4.5% physical damage, and +5% physical damage to champions below 50% HP. For simplicity, we will model the physical damage increase to champions below 50% HP as a consistent 2.5% bonus [8].

In-game modifiers consist of levelling up and purchasing items. Levelling up is accomplished by performing in game actions such as participating in enemy champion kills. Vayne begins the game with 50 AD and 0.568 AS, and gains 3.25 AD and 0.0204 AS per level [7].

Items are purchased with gold earned in game. For our model, we will assume that Vayne has bought a Berserker's Greaves, which grants +0.132 AS [6], and either a BoRK or a BT. A Bloodthirster grants +100 AD, while a BoRK provides + 25 AD, 0.263 AS, and deals bonus physical damage equal to 5% of a target's current health on each hit [5]. For simplicity, we will assume that this is instead consistently 2.5% of a target's HP_{max}.

For our scenario, in which we assume Vayne to be level 9, the cumulative effect of all statistical modifiers except for a BoRK or a BT is 100 AD, 1.01 AS, 6% APen, and +7% physical damage. By adding these values to the k_t formula, we arrive at

$$k_t = \frac{HP_{max}}{1.01(1.07 \times AR_{red} \times 100 + 10 + \frac{HP_{max}}{60})}$$

where AR_{red} is

$$AR_{red} = \frac{100}{100 + (1 - 0.06)AR}$$

Surface Functions
Accounting for the stats given by a BT, the kill time function becomes:

$$k_{tBT} = \frac{HP_{max}}{1.01(1.07 \times AR_{red} \times 205 + 10 + \frac{HP_{max}}{60})}$$

Accounting for the stats given by a BoRK, the kill time function becomes:

$$k_{tBoRK} = \frac{1}{1.27(1.07 \times AR_{red} \times (126 + \frac{HP_{max}}{40}) + 10 + \frac{HP_{max}}{60})}$$

By subtracting the k_t functions, we form a surface function displaying when each item is more effective against a target (figure 1).

$$\Delta k_t = k_{tBT} - k_{tBork}$$

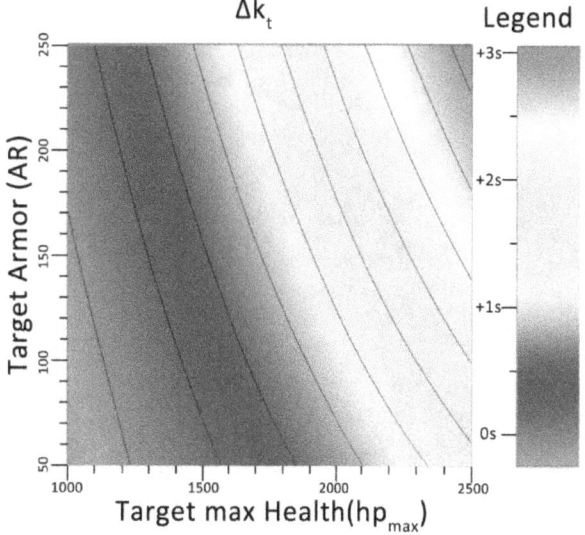

Figure 1 – Δk_t as a function of AR and HP_{max}. Warmer coloured areas indicate that BT has a higher kill time than BoRK, and is therefore less effective. Contours are 1/3 s apart. The bounds were set to values that seem reasonable for the scenario.

Analysis
Purchasing a BT is more effective on targets entering a fight with low HP_{max} (< ~1250) and AR, while buying a BoRK is more efficient on targets with high

HP_{max} and AR. This makes sense, as BoRK deals percentage health damage and makes silver bolts more effective due to the AS it grants. However, the difference in k_t for each item does not appear to be significant, with a maximum difference of about 3.0s seconds between the ideal BT case and the ideal BoRK case.

Conclusion
The difference between building a BT and BoRK does not seem to be appreciable for the scenario described, with a maximum difference of 3 seconds. However, the differences may be more appreciable with more items and levels, and with different runes and masteries.

References

[1] Riot Games, 2014. *Summoner's Rift*. [online]. Available at: http://gameinfo.euw.leagueoflegends.com/en/game-info/game-modes/summoners-rift/ [Accessed 03/13/2014]
[2] Lolking, 2014. *Champion Stats*. [online]. Available at: http://www.lolking.net/champions/ [Accessed 03/13/2014]
[3] LoLWiki, 2013. *V3.10A*. [online]. Available at: http://leagueoflegends.wikia.com/wiki/V3.10a [Accessed 03/13/2014]
[4] LoLWiki, n.d. *Armor*. [online]. Available at: http://leagueoflegends.wikia.com/wiki/Armor [Accessed 03/13/2014]
[5] LoLWiki, n.d. *Attack Damage*. [online]. Available at: http://leagueoflegends.wikia.com/wiki/Attack_damage [Accessed 03/13/2014]
[6] LoLWiki, 2014. *Attack Speed*. [online]. Available at: http://leagueoflegends.wikia.com/wiki/Attack_speed [Accessed 03/13/2014]
[7] LoLWiki, n.d. *Vayne*. [online]. Available at: http://leagueoflegends.wikia.com/wiki/Vayne [Accessed 03/13/2014]
[8] Probuilds, 2014. *Imaqtpie Vayne*. [online]. Available at: http://www.probuilds.net/guide/NA/1290887261/19887289 [Accessed 03/13/2014]

Journal of Interdisciplinary Science Topics

The Monkey King's Somersault

Yannic Chen
The Centre for Interdisciplinary Science, University of Leicester
25/03/2014

Abstract
The Monkey King is a famous fictional character from ancient Chinese literature. A famous saying is that he can travel a distance of 54,000km on a somersault. Such a distance requires a jump height of 13,500km. This feat is inefficient on the Earth, which has dimensions smaller than the jump and as a result, may only be helpful in the Heavenly Kingdom with unknown properties and parameters.

Introduction
The Monkey King, also known as Sun Wukong (or Wukong), is the main character of the novel *"Journey to the West"*, one of the Four Great Classical Novels in ancient China. He is a monkey that was born from a stone and acquired magical powers. One of his more famous powers is the ability to move 108,000 li (里), a distance of 54,000 km, in one somersault. This paper aims to calculate the height, force and duration of such a somersault.

The somersault
The properties of human jump have been of interest to researchers for a long time. Current formulas are very inaccurate and made to fit the data points. A somersault is different from a jump and may be harder to model. As such, for this paper, an easier model is to assume that Sun Wukong moves like a projectile over even ground without air resistance.

Using various equations Sun Wukong's velocity can be found. A projectile travels the farthest when launched at 45° [1]. The distance is 54,000 km and the gravitational constant is 9.81 m/s^2. Putting in the information into the range formula [1], where d is the distance travelled by the projectile, v is the initial velocity, g is the gravitational acceleration and θ is the launch angle:

$$d = \frac{v^2 \sin(2\theta)}{g}$$
$$5.4 \times 10^7 m = \frac{v^2 \sin(90)}{9.81 m/s^2}$$
$$v = 23016 \; m/s$$

After finding the velocity, the duration of the jump can be calculated using the formula below:

$$x(t) = vt\cos(\theta),$$

where $x(t)$ is the horizontal distance travelled in a given time and t is time. The total horizontal distance is $5.4 \times 10^7 m$, the velocity is 23016 m/s and the angle is 45°. From the above formula, rearranging for time, t, gives:

$$t = \frac{x(t)}{v \cos(\theta)}$$
$$t = \frac{5.4 \times 10^7 m}{23016 \frac{m}{s} \times \frac{\sqrt{2}}{2}}$$

$$t = 3318s$$

The maximum height achieved can be calculated by differentiating the vertical displacement equation and equating it to 0:

$$y(t) = vt\sin(\theta) - \frac{1}{2}gt^2$$
$$\frac{dy}{dt} = v\sin(\theta) - gt = 0$$
$$23016\frac{m}{s} \times \frac{\sqrt{2}}{2} - 9.81\frac{m}{s^2}t = 0$$
$$t = 1659s$$

The highest point is reached after 1659 s, which is a height of:

$$y(t) = vt\sin(\theta) - \frac{1}{2}gt^2$$
$$y(1659) = \left(23016\frac{m}{s} \times 1659s \times \frac{\sqrt{2}}{2}\right) - \left(\frac{1}{2} \times 9.81\frac{m}{s^2} \times 2752281s^2\right)$$
$$y(1659) = 1.35 \times 10^7 m$$

The force can be calculated by finding acceleration. In this scenario, Sun Wukong accelerated from 0 m/s to 23016m/s. The acceleration is instantaneous, but since it's not quantitative, we assume the time taken for this acceleration to be 1s. Thus the acceleration will be 23016 m/s^2. The mass of Wukong is not known, thus we can only make assumptions. With a mass of 60kg, the Force will be:

$$F = ma$$
$$F = (60kg)\left(23016\frac{m}{s^2}\right)$$
$$F = 1.38 \times 10^6 N$$

Conclusion

The Journey to the West is said to have happened on Earth in ancient china. The earth has a circumference of 40,075km [2] and an atmospheric height of 10,000km (exosphere) [3]. Although Sun Wukong's somersault is enough to go around the earth once and would be enough to move outside the Earth's atmosphere, the formula did not take into account that the surface is round and neglected air resistance. The force produced is more than 66% of that produced by a space shuttle main engine during lift-off [4]. With such strength, the journey to the east would have been over in just one somersault. One explanation for such feat might be that he only achieved such a distance in the Heavenly Kingdom, assuming that it is much larger than the Earth and has lower gravity, while less force would be needed to travel the same distance, the travel time will be longer.

References

[1] Nave, R. (no date). Trajectories. Retrieved 28. February 2014 from Hyperphysics: http://hyperphysics.phy-astr.gsu.edu/hbase/traj.html
[2] Rosenberg, M. (14. September 2012). Basic Earth Facts. Retrieved 28. February 2014 from About.com: http://geography.about.com/od/learnabouttheearth/a/earthfacts.htm
[3] National Geographic. (no date). Earth's Atmosphere. Retrieved 28. February 2014 from National Geographic: http://science.nationalgeographic.com/science/earth/earths-atmosphere/
[4] Encyclopedia Astronautica. (no date). SSME. Retrieved February 27, 2014, from Astronautix: http://www.astronautix.com/engines/ssme.htm

Journal of Interdisciplinary Science Topics

Breaking Bad: Gus Fring's Face Blown Off

Somaya Turk
The Centre for Interdisciplinary Science, University of Leicester
25/03/2014

Abstract
Be aware that this paper contains spoilers and potentially distressing imagery. Gus (Gustavo Fring) is a fictional character from the American crime drama *Breaking Bad*. In the events of the Season 4 finale, Gus experiences the explosion of a homemade bomb, which kills two characters immediately, but Gus walks out of the room, with half of his face seriously disfigured, and he calmly adjusts his tie before collapsing dead in the doorway. This paper investigates whether his reaction is justifiable, and whether it would occur in real life.

Introduction
Breaking Bad is based on the story of Walt (Walter White), a secondary school Chemistry teacher turned criminal, producing crystal meth (amphetamine) with his partner, Jesse. At one point, Walt attempts to kill the villainous drug lord Gus by planting a homemade bomb under a wheelchair placed in the room where Gus is present with his partner Tyrus and rival Hector.

When the bomb is triggered, the explosion immediately kills Tyrus and Hector; however Gus calmly walks out of the room adjusting his tie, before collapsing dead in the doorway. We are shown Gus's severely disfigured face in the episode, and are left wondering how Gus could react in such an unusual manner [1].

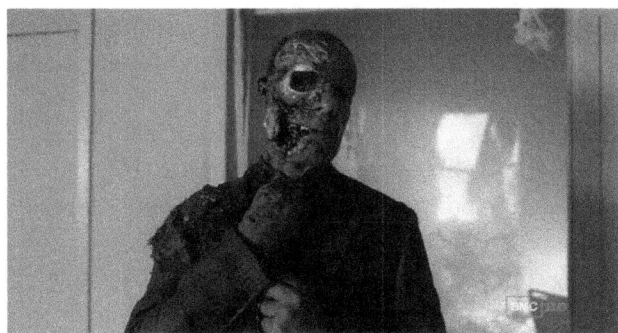

Image 1: Gus Fring's reaction after the explosion [2].

Acute Stress Response
Gus may have been experiencing acute stress response to the explosion, in other words Gus was in a complete state of "shock". When someone goes in to shock, stress hormones such as adrenaline (epinephrine) are released into the bloodstream, and there is increased activity of nerve impulses to various parts of the body. Once the stress response is triggered in the body, a series of changes occur, including: the quickening of pulse, redirection of blood away from extremities and instead to major organs and the release of cortisol which can bring long and short term changes [3], such as Gus' instant response. The elevated cortisol levels would have created physiological changes that helped replenish Gus' body energy, giving him the ability to walk out of the room even for a few brief seconds [4].

Response Initiation
From the episode we can see that the explosion initiates Gus' sensory nerve cells to pass the perception of a threat from the environment to the hypothalamus in his brain. Neurosecretory cells in the hypothalamus transmit signals to the pituitary gland, releasing chemical messengers into his bloodstream. At the same time, the hypothalamus transmits a nerve impulse down the spinal cord. The nerve impulse and chemical messenger both travel to the adrenal gland [5], located above the kidneys [3].

Once the adrenal glands have received these signals, epinephrine is released into the bloodstream. A cell signalling cascade results in the release of cortisol into the bloodstream, where it begins several signalling cascades in various cell types and an increase in glucose levels.

Energy Boost
Once the epinephrine binds to receptors in Gus' liver cells, it triggers a signalling cascade producing glucose from large sugar molecules. Fatty acids are also free to be transformed into energy due to

circulating cortisol. With these molecules circulating in the bloodstream, Gus' muscles are readily provided with energy. This response can also be referred to as the "fight or flight" response [6].

Signals from sensory nerve cells are also received by an area of Gus' brain stem initiating its own responses involving the release of norepinephrine, which again triggers numerous cell signalling cascades in Gus' body. A rapid increase in norepinephrine level provides the body with extra strength and increased arousal, as this particular neurotransmitter is involved with creating several physical reactions and changes such as: muscle tension, increased startle reflex, shortness of breath, dizziness, and numbness and tingling throughout the body [7]. These symptoms are visible when Gus walks out of the room, with an irregular breathing pattern, calm but shaken, and numb to the pain.

Conclusion

By observing Gus Fring's behaviour after the explosion, it is clear to see that his unusual reaction to the explosion was a result of simultaneously released signalling molecules in his body. Combined together, the cell signalling molecules increase the energy and blood circulation to parts of Gus' body which are in most need, preparing him for extreme action – in this case preventing him from feeling any pain for a few brief moments before collapsing to his death.

Furthermore, the injury shows the skull is still intact implying that the motor cortex is not damaged, therefore the fact that Gus walked out of the room adjusting his tie, can be justified by this.

However, looking at the substantial injury that Gus suffered, it is unlikely that he would walk out of the room in such a calm manner, as the sudden release of norepinephrine would make it more likely for Gus to behave erratically. Although we are only shown the top half of Gus' body, and hence can only see half his face and a portion of his torso to be severely disfigured, the explosion may have caused injuries to other parts of his body also. With the additional injuries, his ability to walk would be impaired, leaving him unable to walk if the explosion was to occur in real life.

References

[1] V. Gilligan, Director, Breaking Bad. [Film]. Sony Pictures Television, 2008.
[2] Empire, "Gus Fring's Death," 2013. [Online]. Available: http://www.empireonline.com/features/vince-gilligans-five-best-breaking-bad-moments/p4. [Accessed 14 March 2014].
[3] B. Alberts, A. Johnson, J. Lewis, M. Raff, K. Roberts and P. Walter, "Mechanisms of Cell Communication," in Molecular Biology Of The Cell, New York, Garland Science, 2014, p. 890.
[4] Harvard Health Publications, "Understanding the stress response," March 2011. [Online]. Available: http://www.health.harvard.edu/newsletters/Harvard_Mental_Health_Letter/2011/March/understanding-the-stress-response. [Accessed 24 March 2014].
[5] Genetic Science Learning Center, "How Cells Communicate During Fight of Flight," 2014. [Online]. Available: http://learn.genetics.utah.edu/content/cells/fight_flight/. [Accessed 14 March 2014].
[6] A. Jansen, X. V. Nguyen and T. C. Mettenleiter, "Central Command Neurons of the Sympathetic Nervous System: Basis of the Fight-or-Flight Response," Science, vol. 270, no. 5236, pp. 644-646, 1995.
[7] J. M. Carver, "The Chemical Imbalance in Mental Health Problems," January 2002. [Online]. Available: http://www.drjoecarver.com/clients/49355/File/Chemical%20Imbalance.html. [Accessed 24 March 2014].

Playing 'The Floor is Lava' in Real Life

Deven Fosberry and Pratik Lakhotia
The Centre for Interdisciplinary Science, University of Leicester
28/03/2014

Abstract
The popular children's game 'the floor is lava' seems entertaining when played using only the imagination, but it is not widely known what the effects would be if this game were to be played using real lava. This paper investigates whether playing this game in real life would be possible and what effect that would have on the human body.

Introduction
Most people have played or heard of 'the floor is lava', a popular children's game which requires one to imagine that the floor is made of lava and thus that it cannot be touched. The floor being made of imaginary lava means that one must find alternate methods of traveling from one side of a room to another, resulting in time spent jumping between items of furniture and climbing on various structures in order to avoid an imaginary, yet apparently fiery, death. As an adult, possibly lacking the imagination once possessed as a child, one might start to wonder if it would be possible to recreate this game using real lava in order to reclaim one's youth without putting in the mental effort required to imagine that the living room floor is made of molten rock. This paper seeks to investigate this possibility by taking the many properties of lava into account and determining whether a human would be able to survive in its presence for the duration of the game.

Biological effects of gases produced by lava
Assuming that the floor being used for this game has spontaneously transformed into lava, rather than arriving via an erupting volcano, we can also assume that the huge concentrations of gases produced during a volcanic eruption would not be present in the atmosphere. However, the lava itself would likely produce gases of its own, independent of any volcano.

The main gases produced by lava are water vapour and carbon dioxide (CO_2) [1], the latter of which could be quite dangerous if we were to play this game in an area with an uneven floor. This is because CO_2 is heavier than air so it would displace the air and accumulate in the lowest points of the room. In these areas, a person would experience a rapid loss of consciousness, followed by asphyxiation due to the lack of oxygen [2]. People in areas with lower concentrations of CO_2 might experience difficulty breathing, dizziness, and impaired coordination before eventually losing consciousness [3], all of which would make playing this game quite difficult.

Examples of other gases which are produced in minor amounts by lava are sulphur dioxide (SO_2), hydrogen sulphide (H_2S), and occasionally hydrogen fluoride (HF) [3]. In small amounts, these gases can cause irritation of the mucous membranes of the eyes, nose, and throat. Long exposure to low concentrations of these gases can cause similar breathing issues to those experienced with lower concentrations of CO_2 as well as dizziness and irritation of the respiratory tract. Exposure to HF can be especially dangerous as it is a strong irritant and larger concentrations of it can be highly poisonous [3]. These gases are not usually produced in large amounts from lava alone, though, so as long as the game being played does not last for several hours, it may be possible to avoid the most serious of these effects.

Air temperature above lava
In order to see what effect the extreme temperatures of lava would have on the human body, we can calculate the temperature of the air above the lava in the house that the game would be played in. To do this we use the following equation:

$$Energy\ lost\ by\ lava = energy\ gained\ by\ air$$
$$-m_{lava}c_{lava}(T_i - T_f) = m_{air}c_{air}(T_f - T^i) \quad (1)$$

Where m is mass in kilograms, c is the specific heat capacity (in J kg^{-1} K^{-1} – 0.84 for lava and 1.01 for air

[4]), T_i is the initial temperature in Kelvin (1523K for lava [5] and 293K for air [6]), and T_f is the final temperature in Kelvins. The assumptions that we need to make in order to use this equation are that the house is insulated to the extent that it becomes a closed system (meaning no heat can escape), the floor is made of a common type of lava, basalt, to a depth of 0.5m, and that the room being used for the game is 3m tall with a floor area of 1m². We also need to calculate the respective masses of lava and air present in the room at the time of the game, which is done by multiplying their densities by their volumes. The density of lava is $2700 \, kg \, m^{-3}$ [7] which gives it a mass of $1.35 \times 10^3 \, kg$, and the density of air is $1.204 \, kg \, m^{-3}$, giving it a mass of $3.612 \, kg$.

Using these values, we can solve equation (1) to give us a final temperature of 1519K, which would also be the temperature of any room with the same height and depth of lava but different floor area. This tells us that the game could not actually be played indoors as the temperature would be too high for a human to tolerate. Because of this fact, we may want to consider the possibility of playing this game outdoors, under the assumption that a more open space would cause the temperatures above the lava to be much lower, allowing the players to actually survive.

To calculate the rate of temperature change above the lava while playing the game outdoors, we can use the Stefan-Boltzmann law:

$$P = \epsilon \sigma A (T^4 - T_c^4) \quad (2)$$

In this equation, P is equal to power in Watts, ϵ is the emissivity (0.9 for lava [8]), A is the area above the lava, σ is $5.67 \times 10^{-8} W \, m^{-2} \, K^{-4}$, T is the temperature of the lava (1523K), and T_c is the temperature of the surroundings. We also assume that the furniture we are stood on has a height of 1m and that the lava is contained within an area of 1m². This gives us a value for P of $2.75 \times 10^5 \, W$ which is equal to $2.75 \times 10^5 \, J \, s^{-1}$.

The next step is to calculate the change in temperature per second above the lava, and thus determine how long one would be able to play the game for. We know that:

$$Energy \, per \, second = mc\Delta T \, per \, second \quad (3)$$

So, assuming that the mass of the player is 70kg, we can use this to calculate a change in temperature (ΔT) of $3.89 \, K \, s^{-1}$. This tells us that it would only be a matter of seconds before the temperature increases to such a point that the game is interrupted by the death of the players.

However, these calculations do not account for convection currents, which would play a large role as the participants are moving around directly above the lava. The existence of convection currents means that there would be a much higher temperature change above the lava than predicted here. The convection currents also account for the large number of photographs of geologists and rock enthusiasts who are stood relatively close to lava as the heat dissipates quickly near the edges.

Conclusion

In conclusion, from our calculations, we can see that it would not be possible to play a game of 'the floor is lava' either indoors or outdoors for more than a few seconds due to the large air temperatures produced above the lava. However, the authors of this paper do believe that it may be possible to play a short game of 'the walls are lava' if convection currents are taken into account, as long as the room it is being played in does not have a ceiling and participants attempt to inhale as few of the gases as possible.

References

[1] Nelson, Stephen A. (2013). *Characteristics of Magma.* Available: http://www.tulane.edu/~sanelson/Natural_Disasters/volcan&magma.htm. Last accessed 6th Mar 2014.
[2] Rice, Susan A. (2004). *Human health risk assessment of CO_2: survivors of acute high-level exposure and populations sensitive to prolonged low-level exposure.* Available: http://www.netl.doe.gov/publications/proceedings/04/carbon-seq/169.pdf. Last accessed 6th Mar 2014.

[3] USGS. (2010). *Volcanic Gases and Their Effects.* Available: http://volcanoes.usgs.gov/hazards/gas/. Last accessed 7th Mar 2014.
[4] Engineering Toolbox. (). Solids – Specific Heats. Available: http://www.engineeringtoolbox.com/specific-heat-solids-d_154.html . Last accessed 6th March 2014.
[5] Oregon State University. (). *How hot is lava?*. Available: http://volcano.oregonstate.edu/how-hot-lava. Last accessed 6th March 2014.
[6] Engineering Toolbox. (). Air Properties. Available: http://www.engineeringtoolbox.com/air-properties-d_156.html. Last accessed 6th March 2014.
[7] Murase, Tsutomu and McBirney, Alexander R. (1973). (Properties of Some Common Igneous Rocks and Their Melts at High Temperatures). Available: http://bulletin.geoscienceworld.org.ezproxy3.lib.le.ac.uk/content/84/11/3563.full.pdf+html. Last accessed 6th Mar 2014.
[8] Del Negro, C., Fortuna, L., Herault, A., Vicari, A . (). *SIMULATIONS OF THE 2004 LAVA FLOW AT ETNA VOLCANO BY THE MAGFLOW CELLULAR AUTOMATA MODEL* . Available: http://www.earth-prints.org/bitstream/2122/2627/1/1368.pdf. Last accessed 6th Mar 2014.

Journal of Interdisciplinary Science Topics

Can we power a spaceship?

Somaya Turk, Pratik Lakhotia & Deven Fosberry
The Centre for Interdisciplinary Science, University of Leicester
28/03/2014

Abstract
This paper discusses the possibility of humans being able to produce enough energy, via effective exercising, in order to power a single space ship. The dimensions of the spaceship were taken using a paper model of the Axiom spaceship from the film WALL-E, and were scaled up to store the population value obtained from the movie. It was concluded that the acceleration of the Axiom in space would be 1.14 ms^{-1} per day, if powered by a population of exercising adults.

Introduction
In the Pixar animation film WALL-E, the Axiom (a spaceship) contains a population of 600,000 people [1]. This paper talks about the possibility of powering the spaceship, using the energy provided by the humans doing exercise rather than its usual source of power.

Modelling the Axiom spaceship
The Oasis of the Seas is the modern day equivalent of the Axiom as it's the world's largest cruise ship so we assume that they have a similar density. This was then used to calculate mass of the Axiom.

The Axiom was scaled up as a cylinder and the following values were as used, as demonstrated in table 1:

	Oasis of the sea	Axiom
Length (m)	362	4540 [2]
Beam* width (m)	625	1188 [3]
Beam radius (m)	312.5	594
Volume (m^3)	1.44x10^5	8.78x10^6
Mass (kg)	1x10^8	6.10x10^9
Density (Kgm^{-3})	694	694

Beam*: The width at the widest point.
Table 1: Shows a comparison between the cruise ship and the Axiom [4, 5].

Image 1: Shows a comparison between the Oasis of the Sea (left), and the Axiom (right) [3].

The Energy Produced By Humans
A number of assumptions were made when calculating the dimensions of the spaceship, and the energy produced by each adult on the Axiom. Looking at Image 2, we can assume that the overweight population on the Axiom weigh 160kg on average; in comparison to the average North American (highest body mass of any continent) weighing 80.7kg [6], moderate exercise on an exercise bike burns around 768 calories an hour [7] which is equivalent to 3379.2 joules.

Image 2: Image showing the appearance of the humans in the movie, from this we can assume their weight to be much higher than the average human weight [8].

We know that there is a population of 600,000 and if each person does 2 hours of exercise a day:

3379.2 J hr^{-1} x 2 hr x 600,000 = 4.06x10^9 Joules

The exercise would not necessarily have to be done in one go, as this could possibly cause major health issues. Instead, it should be sufficient to assume that the 2 hours of exercise is completed over the course of one day. In addition, the humans would need to sustain their bodies, and hence they would be required to consume enough calories, using the resources available on the Axiom, to maintain their body mass.

We can then convert this energy to velocity using the equation for kinetic energy. The mass used will need to also include that of the population of the Axiom:

$$m = 6.10 \times 10^9 + (160 \times 600000)$$
$$= 6.20 \times 10^9 kg$$

$$E = \frac{1}{2}mv^2$$

$$v = \sqrt{\frac{2E}{m}} = \sqrt{\frac{2 \times 4.06 \times 10^9}{6.20 \times 10^9}} = 1.14 \, ms^{-1}$$

For this calculation, the deceleration is negligible due to the lack of a medium in space. This value would be the acceleration per day, assuming that every person carries out 2 hours of exercise in each 24 hour period.

$$Time = \frac{Velocity}{Acceleration} = \frac{c}{a} = \frac{3 \times 10^8 ms^{-1}}{1.14 ms^{-1} day^{-1}}$$

$$= 2.63 \times 10^8 days = 720{,}000 \, years$$

Assuming relativistic effects do not take place, it will take 720,000 years for the Axiom to reach the speed of light.

In the movie, the Axiom only travelled for 700 years [1], which means that it would only reach a speed of 2.91 x 10^5 m s^{-1}.

Conclusion

Assuming that the humans are the ship's only power source, operating with 100% efficiency, as well as the other assumptions stated this paper, we can see that humans create 4.06 x 10^9 Joules of energy via medium exercise. This equates to an acceleration of 1.14 ms^{-1} per day on-board the Axiom. With this acceleration, the axiom would take over 700,000 years to accelerate to the speed of light but, in the 700 years it was away from the Earth in the movie, it would only reach a speed of 2.91 x 10^5 m s^{-1}. From this, we can conclude that it is indeed possible for humans to power a spaceship through exercise. However, its velocity may not be ideal so it may be wise to seek out alternative power sources.

References

[1] Pixar wiki. (2008). *Axiom.* Available: http://pixar.wikia.com/Axiom. Last accessed 14th Feb 2014.
[2] Dirkloechel. (2013). *Size Comparison - Science Fiction Spaceships.* Available: http://dirkloechel.deviantart.com/art/Size-Comparison-Science-Fiction-spaceships-398790051. Last accessed 14th Feb 2014.
[3] Perdana, Julius. (2008). *The AXIOM Spaceship WALL-E Paper Model.* Available: http://paper-replika.com/index.php?option=com_content&id=59. Last accessed 14th Feb 2014.
[4] Royal Caribbean International. (2013). *Oasis of the Seas.* Available: http://www.royalcaribbean.co.uk/our-ships/oasis-class/oasis-of-the-seas/ships-decks-and-facts/#tab-16-4295. Last accessed 14th Feb 2014.
[5] Boston.com. (2006). If Royal Caribbean builds it, 6,400 could come. Available: http://www.boston.com/travel/getaways/us/florida/articles/2006/02/07/if_royal_caribbean_builds_it_6400_could_come/
[6] Walpole, S; Prieto-Merino, D; Edwards, P; Cleland, J; Stevens, G and Roberts, I. (2012). The weight of nations: an estimation of adult human biomass. Available: http://www.biomedcentral.com/1471-2458/12/439
[7] British Heart Foundation. (2014). *Exercise Calorie Calculator.* Available: http://www.bhf.org.uk/heart-health/prevention/calorie-calculator.aspx. Last accessed 14th Feb 2014.
[8] D. Wogan, "Scientific American," 2012. [Online]. Available: http://blogs.scientificamerican.com/plugged-in/2012/05/17/approaching-wall-e-with-hondas-uni-cub-personal-mobility-device/. [Accessed 24 March 2014].

Journal of Interdisciplinary Science Topics

The Curious Case of the Glowing Bones

Stephanie Taylor
The Centre for Interdisciplinary Science, University of Leicester
28/03/2014

Abstract
In the FOX television series *Bones* a set of remains are discovered that are blue and glowing. It was determined that the bones were glowing due to the presence of a *Vibrio phosphoreum*, which is a bioluminescent bacteria. This paper looks into whether or not the *V. phosphoreum* could have been introduced into the victim's body and survived long enough to be present when the remains were found.

Introduction
In the FOX television series *Bones* each episode introduces a new crime scene where the remains are unrecognizable. In order to solve the crimes, Dr. Temperance Brennan, a forensic anthropologist, and her team are called in to analyse the remains. Throughout the duration of the show, there have been some very unusual crime scenes. One of the most unusual occurred in Season 2, Episode 20: *The Glowing Bones in the Old Stone House* [1]. In this episode, the bones were blue and glowing. It was determined that this was a result of the introduction of the bacteria *Vibrio phosphoreum* into the victims blood stream through a cut made by a knife with the bacteria on it. This seems like a very unlikely scenario, and it begs the question could this actually happen?

Episode Background
In this episode Dr. Brennan and her team are called in to identify remains found in an old stone house [1]. After further analysis, they determined that the bones were covered in *V. phosphoreum*, which is commonly found in marine organisms such as squid, shrimp, and sea urchins. The victim was a very popular chef who had been learning how to make sushi when she was cut with a knife, introducing the bacteria into her blood stream. It was also believed that the presence of *V. phosphoreum* aided the speedy decomposition of the victim.

After she was murdered, the *V. phosphoreum* in her blood stream continued to live in her body, and as her body decayed, it ended up coating her exposed bones [1]. To determine if this is possible, two things must be determined: could the *V. phosphoreum* live in the human body and could it continue living long enough to speed the decay of the body and be found three days later, still glowing?

V. phosphoreum in the Human Body
There are many strains of bacteria that are bioluminescent [2]. *V. phosphoreum* is the strain that emits the strongest bioluminescent glow. It can easily be isolated from fish and other marine organisms, allowing it to be easily transferred to new organisms. However, it has been classified as a marine organism because it requires sodium concentrations of 200mM or greater in its growth medium [3]. The average blood sodium level in humans is 135 – 145mM [4]. This shows that it is unlikely *V. phosphoreum* could survive in the human blood stream, unless the victim has an above average blood sodium level.

In addition *V. phosphoreum* has an ideal growth temperature of 4°C, which is much lower than human body temperature. In fact it has been found that *V. phosphoreum* cannot grow at temperatures higher than 35°C [3]. This means that there was no way that *V. phosphoreum* could survive long in the human body, which has an average temperature of 36 – 37°C [5]. Since the victim was cut a few days before her death, and it was three more days after her death that the remains were discovered, it is extremely unlikely that the *V. phosphoreum* could have survived in her body that long.

A Different Bioluminescent Bacterium
While it does not look as though *V. phosphoreum* could have been the bacteria covering the remains, that does not entirely eliminate the possibility that the bones could have been covered in bioluminescent bacteria. There are multiple other

strains of bioluminescent bacteria, which are pathogenic and can live within the human body [2].

One such strain is *Vibrio fischeri*, which is a member of the same genera as *V. phosphoreum*. It shares many of the same characteristics as *V. phosphoreum* but its differences may make it a better candidate for the bacteria used in *The Glowing Bones in the Old Stone House*.

It can be found free living in marine environments or associated with fish [2, 6]. *V. fischeri* cannot grow at 4°C, but it thrives at 35°C, which is approximately human body temperature [2]. This indicates that *V. fischeri* could have been transferred from a fish used in sushi to the victim by the knife. In addition *V. fischeri* is often found with Bobtail squid, which can be used in sushi [6]. This supports the story of a sushi knife being used to introduce the bacteria into the victim's bloodstream.

V. fischeri is a marine organism, but it has been found in fresh water environments [6]. This shows that while it will thrive in environments with high levels of sodium ions, it can exist in environments with sodium levels as low as 0.87mM [7]. This indicates that it is possible for *V. fischeri* to survive in the bloodstream, sodium levels are much higher than 0.87mM.

V. fischeri is also associated with human disease (especially blood poisoning) caused by consumption of contaminated seafood and exposure of wounds to contaminated sea water [6]. This further supports that *V. fischeri* could have survived in the victim's bloodstream. Since *V. fischeri* is pathogenic, it is very likely that the victim would have contracted blood poisoning after her cut. But, as she died within days of her cut, it is unlikely that the symptoms had manifested to an extent that they would have been recognized. Why the possible blood poisoning did not show up on the autopsy is another question entirely.

Finally, *V. fischeri* also releases blue-green light, which is similar to the colour of light emitted by the bones in *The Glowing Bones in the Old Stone House* [6]. In most cases the light emitted by *P. fischeri* is not that bright, which means that it may not have been as visible as the bacteria on the bones [2].

Conclusion
When looking into *V. phosphoreum* further, *V. phosphoreum* could not exist in the human body. There is a chance that the exact species of bacteria was simply misidentified by the members of Dr. Brennan's team. The bacteria species *V. fischeri*, which is in the same genera as *V. phosphoreum* could theoretically survive in the human body.

Further investigation is needed to determine if either of the bacteria discussed could have covered the bones. In addition, the bones were discovered bare after only three days, and while this may have been due in a large part to scavengers present, there is a chance that the bacteria could have also contributed.

References

[1] Anon, 2014. *The Glowing Bones in the Old Stone House*. [online] Bones Wiki. Available at: <http://bones.wikia.com/wiki/The_Glowing_Bones_in_the_Old_Stone_House> [Accessed 14 Mar. 2014].
[2] Eddleman, H., 1999. *Vibrio phosphoreum from Squid or Ocean Fish*. Available at: <http://www.disknet.com/indiana_biolab/b203.htm> [Accessed 14 Mar. 2014].
[3] Reichelt, J. and Baumann, P., 1973. Taxonomy of the Marine, Luminous Bacteria. *Arch. Mikrobiol.*, **94**, pp.283–330.
[4] U.S. National Library of Medicine, 2014. *Sodium - blood*. [online] Medline Plus. Available at: <http://www.nlm.nih.gov/medlineplus/ency/article/003481.htm> [Accessed 14 Mar. 2014].
[5] NHS Direct, 2008. *About high temperature*. Available at: <https://www.nhsdirect.nhs.uk/~/media/Selfcare/ColdsAndFlu/HighTempWithLinks.ashx> [Accessed 14 Mar. 2014].
[6] Kenyon College, 2011. *Vibrio fischeri*. [online] Microbe Wiki. Available at: <http://microbewiki.kenyon.edu/index.php/Vibrio_fischeri> [Accessed 14 Mar. 2014].

[7] US EPA, O., 2014. *Sodium in Drinking Water*. [online] Available at: <http://water.epa.gov/scitech/drinkingwater/dws/ccl/sodium.cfm> [Accessed 14 Mar. 2014].

Katniss's Flaming Wedding Dress in Real Life

Radvile Soryte
The Centre for Interdisciplinary Science, University of Leicester
31/03/2014

Abstract
This paper discusses the feasibility of the fire dress from "The Hunger Games: Catching Fire" in real life. It was decided that the dress could be made with two layers: one upper layer composed of cotton and one lower layer composed of a fire-resistant fabric, Kevlar The paper calculates the thermal current between the materials and determines the temperature transferred to the skin. Hence, it was concluded that the flaming dress could be worn for 5 seconds maximum and only under extra precautions.

Introduction
The latest sequel of "The Hunger Games" movie stunned thousands of women around the world with a fabulous flaming wedding dress, which was worn by Katniss during a television interview before the start of the Hunger Games (figure 1).

Figure 1. Katniss's fire dress [1]

As Katniss spun around, the dress caught fire, causing the upper white fabric to burn away and reveal the black dress underneath. It looked amazing in a movie, but would it be possible to create such a dress in real life? To achieve this, this paper tries to model a dress that would be made from two layers of different fabrics - Kevlar, fire insulating material, and cotton, known as the fastest burning fabric [2].

Designing the fire dress
Kevlar was chosen as the fire insulating material and hence placed beneath the cotton fabric to protect the skin from the fire. Kevlar is a unique material, because it does not melt, but degrades at relatively high temperatures (427°C - 482°C) [3]. Hence, in order to achieve a similar effect to the one seen in the movie Kevlar would need to be under the cotton layer that would be set on fire. However, in real life the heat from the fire might be unbearable for a human, even with the insulating material. Therefore, by calculating the thermal current between the materials it is possible to determine the amount of heat transferred to the surface of skin through the insulating material.

How hot it would get?
In order to determine the amount of heat produced by burning the dress, we can rearrange an equation for the thermal current [4]:

$$I = \frac{dQ}{dt} = -kA\frac{dT}{dx}$$

If we solve this equation for the temperature difference, we obtain:

$$\Delta T = I\frac{\Delta x}{kA}$$

which can be simplified to:

$$\Delta T = IR$$

Hence, for the thermal current we get:

$$I = \frac{\Delta T}{R}$$

where I represents thermal current, R is thermal resistance and ΔT is the change of temperature. Therefore, we need to find the thermal resistances of both Kevlar and skin. They can be calculated by applying the equation of the thermal resistance is:

$$R = \frac{x}{kA}$$

where x is thickness of material, A is area and k is thermal conductivity. An average wedding dress requires approximately 18m² of fabric [5], meaning we would need the same amount of Kevlar under the upper layer of the dress. Also, the average thermal conductivity of Kevlar is 0.06 W/mK [6] and the assumed thickness of Kevlar is 0.03m. Hence,

$$R(kevlar) = \frac{0.03m}{0.06W/mK \times 18m^2} = 0.027 m^2 K/W$$

In order to calculate thermal resistance for skin we know that the area of human skin is approximately 2m², thickness is 0.003m and thermal conductivity is 0.37 W/mK [7]. Hence, the thermal resistance is:

$$R(skin) = \frac{0.003m}{0.37W/mK \times 18m^2} = 0.0041 m^2 K/W$$

Now that we have the thermal resistance of Kevlar and skin we can determine the thermal current of the burning dress to see how much heat would be transferred to the surface of skin through the insulating material. Hence, in this case the equation would be:

$$I = \frac{\Delta T}{R(kevlar) + R(skin)}$$

Also, the burning point of the cotton is 210°C [8] and we know that the normal temperature of human body is 36.6°C. Hence,

$$I = \frac{210°C - 36.6°C}{0.027 m^2 K/W + 0.0041 m^2 K/W} = 5575.5\ W$$

Then we can calculate the temperature difference across the Kevlar [4]:

$$\Delta T = IR(kevlar) = 5575.5W \times 0.027 m^2 KW = 150.54°C$$

Now we are able to determine the temperature at the interface between Kevlar and the skin:

$$T = 210°C - 150.54°C = 59.45°C$$

According to the American Burn Association 60°C hot water can cause a third degree burn after 5 seconds [9]. Meaning, that the safest amount of time to wear a flaming dress would be maximum 5 seconds.

Conclusion
Knowing that water as hot as 60°C can cause severe burns after 5 seconds, it can be concluded that a flaming dress could be worn for a maximum of 5 seconds. However, this paper did not consider the fact that the dress is sleeveless so flames could easily the harm hands, neck and other body parts that are not protected from the fire. Also, as a dress would require at least 3cm of Kevlar, most women would not wear it, because it would make them look at least 2 sizes bigger. As a result, a flaming dress like the one described would be suitable only for adrenaline fans that would like to try the dress only for the purpose of science or fun.

References:

[1] Katniss' Wedding Dress And 8 More Futuristic Movie Fashions We Want In Our Closets | MTV.com. 2014. *Katniss' Wedding Dress And 8 More Futuristic Movie Fashions We Want In Our Closets | MTV.com*. [ONLINE] Available at: http://www.mtv.com/news/articles/1719957/futuristic-movie-fashions-we-want.jhtml. [Accessed 21 February 2014].

[2] Burning Characteristics of Fibers. 2014. *Burning Characteristics of Fibers*. [ONLINE] Available at: http://missourifamilies.org/features/materialarticles/feature7.htm. [Accessed 13 March 2014].

[3] Kevlar, 2014. Kevlar Technical Guide. [ONLINE] Available at: http://www2.dupont.com/Kevlar/en_US/assets/downloads/KEVLAR_Technical_Guide.pdf . [Accessed 13 March 2014].

[4] Tipler, P.A., 2008. *Physics For Scientists and Engineers*. 5th ed. 675-679p.

[5] After the dress…: The "How Much Fabric?" Series: Dresses. 2014. *After the dress…: The "How Much Fabric?" Series: Dresses*. [ONLINE] Available at: http://afterthedress.blogspot.co.uk/2009/06/how-much-fabric-series-dresses.html. [Accessed 14 March 2014].

[6] Vettori, R., 2005. Estimates of Thermal Conductivity for Unconditioned and Conditioned Materials Used in Fire Fighters' Protective Clothing. *Building and Fire Research Laboratory National Institute of Standartds and Technology*, 20899-8661, 33.

[7] Anderson GS., Martin AD., 1994. Calculated thermal conductivities and heat flux in man. *Undersea Hyperb Med.*, 431-41.

[8] Cotton - Wikipedia, the free encyclopedia. 2014. *Cotton - Wikipedia, the free encyclopedia*. [ONLINE] Available at: http://en.wikipedia.org/wiki/Cotton. [Accessed 14 March 2014].

[9] American Burn Association, Scald Injury Prevention Educator's Guide. [ONLINE] Available at: http://www.ameriburn.org/Preven/ScaldInjuryEducator%27sGuide.pdf . Accessed 13 March 2014].

Journal of Interdisciplinary Science Topics

Is Dr Conner's Regenerative Transformation Possible?

Vincent So
Honours Integrated Science Program, McMaster University
31/03/2014

Abstract
In *The Amazing Spiderman*, Dr Connors injects himself with a regenerative serum containing lizard DNA which causes an amputated arm to regrow overnight. This article explores the possibility of utilizing signal molecules, found in both humans during embryotic development and the healing response of animals with high regenerative capacity, for the regeneration of human limbs. The treatment is found to be plausible but the timeframe for the regrowth of Connor's arm is unrealistic.

Introduction
In *the Amazing Spiderman,* Peter Parker visits Dr. Curt Connors, a scientist who studies the regenerative properties of lizards [1]. During this visit Parker provides Connors with the missing algorithm necessary for his regenerative serum [1]. Soon after, Connors injects himself with a lizard DNA serum and regenerates his amputated arm overnight [1]. Unfortunately, the serum had unintended side effects and Connors mutates into an evil lizard [1]. Although this scenario is fictional, one is left to wonder if it is scientifically possible. The story poses a real question for regenerative medicine; is it feasible for humans to regenerate missing limbs?

Limb Regeneration in Nature
Humans have a natural healing capacity, although the human regenerative response is comparatively limited [2]. Although mammals lack the ability to regenerate limbs, they are capable of undergoing healing for damaged skin, muscles, bone tissue, tendons and ligaments [2, 3]. Urodeles (an order of amphibians) and zebrafish are typically used to study limb regenerative capacity, providing valuable insight [4]. The limb regeneration process typically occurs in three steps — wound healing, dedifferentiation, and redevelopment (as seen in Figure 1) [4]. Wound healing involves the formation of an epithelial cell layer over the amputation plane and the initiation of the inflammatory response [2]. Dedifferentiation involves the induction of regeneration-specific genes, and the formation of an apical epithelia cap (AEC) [4]. Dedifferentiation also involves the creation of blastema (bs); a mass of undifferentiated stem cells that can undergo redevelopment [3]. Redevelopment involves the

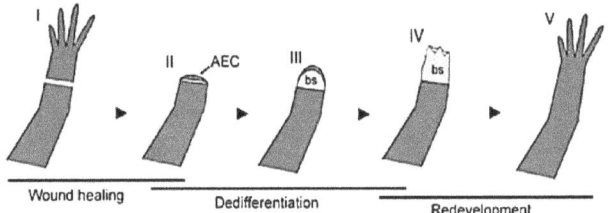

Figure 2: Steps of Limb Regeneration including wound healing, dedifferentiation, and redevelopment [4].

growth of the blastema, creating new limbs distal to its site of origin and re-differentiation of cells. This process also involves close interactions between the blastema and AEC, likely mediated by growth factors such as fibroblast growth factors (FGFs), used in limb elongation during embryonic development [4].

Signals Involved in Initiation of Limb Regeneration
Although limb regeneration is a three-step process, many believe that the signal molecules associated with the early processes ultimately decide the final outcome [4]. Some of these signal molecules include matrix metalloproteinases (MMPs), fibroblast growth factors (FGFs), and molecules of the Wnt/Beta-catenin pathway [3, 4]. MMPs are enzymes activated during inflammation and wound healing and are also unregulated early on in limb regeneration [4]. Other molecules vital to the initiation process are Wnt proteins and Beta catenin is Wnt/ β-catenin, which are proteins that control cell proliferation through the regulation of target gene expression [4]. In fact, when organisms were treated with a MMP inhibitor or a Wnt/ β-catenin inhibitor following amputation, limb regeneration was partially inhibited [4]. The Wnt/Beta-catenin pathway also induces the expression of the *fgf-10* gene. These FGF molecules maintain interactions between the ACE and the blastema, necessary for limb outgrowth [4]. Studies show that two *fgf* genes

are expressed after the amputating of a regenerative limb and neither is expressed after amputating a non-regenerative limb [4]. Moreover, it was shown that when treated with exogenous β-catenin or FGF-10, the regenerative capacity in non-regenerative limbs was improved [4].

Was Dr. Curt Connor's Transformation Feasible?
First of all, is it possible for humans to undergo limb regeneration despite a relatively low regenerative potential? The fact that the key genes expressed during limb regeneration in animals with high regenerative potential are identical to the genes expressed during human development supports the idea that humans possess the appropriate genetic machinery [5]. It has been suggested that by reinforcing the positive feedback loop between signal molecules such as *fgf* genes, the regenerative capabilities of non-regenerative vertebrate limbs may be enhanced [4]. If the serum used allowed the human body to access mechanisms associated with certain signal molecules, limb regeneration would be possible. An important second question to ask is whether or not Connor's transformation occurred in a realistic timeframe. To perform such a calculation, one must assume that there is a constant growth rate and is dependent on the length of the regenerate limb (distance) and time. The equation for the rate of growth is therefore:

$$\text{Rate of Growth} = \frac{\text{Length of Limb to Regenerate}}{\text{Time}}$$

With an associated standard error defined as:

$$\Delta Z = Z \sqrt{\left(\frac{\Delta A}{A}\right)^2 + \left(\frac{\Delta B}{B}\right)^2}$$

where Z is equal to rate of growth, A corresponds to the length of limb to regenerate, B corresponds to the time, and the ΔZ, ΔA, and ΔB correspond to the associated errors. For this calculation, time is set to 8 ± 2 hours, assuming normal sleep patterns. Based on Figure 2 and the assumption that his arm is approximately 0.75 m long, the length of the amputated arm is estimated to be 0.50 ± 0.02 m long. Therefore, the length of regenerate limb would simply be the difference between the arm length and the amputated arm. Using these values, Connor's growth rate is estimated to be 1500 ± 380 mm/day. Although Connors utilizes lizard DNA in his

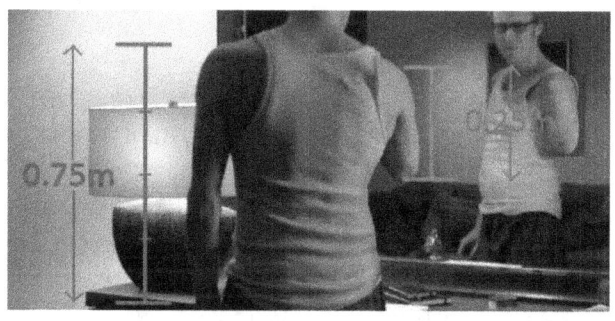

Figure 2: A modified image depicting Dr. Curt Connor's amputated arm and the associated length [4].

serum, his transformation was compared to that of a zebrafish because of the nature of the serum [6]. Due to the fact that there is a lack of well-developed molecular and genetic methods for manipulating genes in amphibians, they are typically not used in experiments involving signal molecules [6]. However, zebrafish are amenable to standard molecular and genetic manipulations such as mutagenesis screens, where specific genes necessary for limb regeneration have been identified [7]. In addition, the zebrafish has one of the fastest regrowth rates in nature, so it sets the relative limit for rates found in nature [6]. This comparison is also more effective because it compares Connor's transformation to organisms shown to express similar genes that would theoretically be used in a serum [6]. When compared to the growth rate of a zebrafish (0.1 to 0.4 mm/day) following fin amputation, the rate of growth for Connor's seems highly unrealistic [7].

Conclusion
In essence, limb regeneration seen in the movie *the Amazing Spiderman* is a very real possibility in the nearby future assuming that such a gene therapy passes necessary ethical and medical criteria. Such treatment would aim to enhance the inherent regenerative abilities of the human genome and associated cellular mechanisms, rather than creating a human-animal hybrid. In addition, even if a functional regenerative serum were made, the growth rate of 1500 ± 380 mm/day seen in the film would not be feasible. So, although the side effects of a future limb regeneration serum may hold equally dangerous results, no one needs to worry about a lizard villain with regenerative capabilities roaming the streets.

References

[1] Columbia Pictures, Marvel Entertainment, 2012. *The Amazing Spiderman* (film).
[2] Han, M., Xiaodong, Y., Taylor, G., Burdsal, C.A, Anderson, R.A., and Muneoka, K., 2005. *Limb Regeneration in Higher Vertebrates: Developing a Roadmap*, The Anatomical Record **287B,** 14-24.
[3] Kumar, A., Godwin, J.W., Gates, P.B., and Brockes, J.P., 2014. *Molecular Basis for the Nerve Dependence of Limb Regeneration in an Adult Vertebrate,* Science **318**, 772-777.
[4] Yokoyama, H., 2008. *Initiation of limb regeneration: The critical steps for regenerative capacity*, Develop. Growth Diff. **50**, 13-22.
[5] Marvel, n.d., Curtis Connors, Available at: http://marvel-movies.wikia.com/wiki/Curtis_Connors.
[6] Poss, K.D, Keating, M.T, Nechiporuk, A., 2003. *Tales of Regeneration in Zebrafish*, Developmental Dynamics **226**, 202-210.
[7] Lee, Y., Grill, S., Sanchez, A., Murphy-Ryan, M., Poss, K.D, 2005. *Fgf signalling instructs position-dependent growth rate during zebrafish fin regeneration,* Development **132,** 5173-5183.

Journal of Interdisciplinary Science Topics

Powers of Poison: The Science Behind Sherlock

Nicole Lindsay-Mosher and Katie Maloney
Integrated Science, McMaster University
04/04/2014

Abstract
In the popular BBC television series *Sherlock*, the villainous mastermind Moriarty commits a murder by introducing botulinum toxin to a victim's eczema cream. This paper examines the science behind this scenario and analysis indicates that it is feasible to commit murder in this manner but a longer timeframe than described in Sherlock is required.

Introduction
The BBC television program *Sherlock* is a modern-day adaptation of the famous book series by Sir Arthur Conan Doyle, which follows the adventures of the fictional detective Sherlock Holmes and his colleague Dr. John Watson. During the episode "The Great Game", Holmes must solve the murder of a child named Carl Powers. Holmes examines the victim's sneakers and finds traces of botulinum toxin, a deadly neurotoxin.

Carl Powers
The victim of the botulinum toxin was an 11 year old boy named Carl Powers. He was a champion swimmer from Sussex who was visiting London for a swimming competition when he mysteriously drowned. Powers was known to have eczema and the only evidence of the toxin was found in his shoes indicating that the toxin had been introduced through his eczema cream. Holmes deduces that Powers had applied the toxin-laced medication two hours prior to his swimming competition, at which point the poison took effect and paralysed his muscles, causing him to drown. The two-hour delay was inferred based on the time it would have taken for Powers to travel from his hometown of Sussex to his competition in London. Holmes guesses that the murderer, Moriarty, had strategically introduced the toxin right before Powers' swimming race, causing him to drown in the pool.

Absorption of Botulinum toxin
Botulinum toxin is a potent neurotoxin produced by the bacterium *Clostridium botulinum* [1]. Botulinum toxin cannot be absorbed through intact skin; however, it can enter the bloodstream through open wounds or lesions [2]. Powers had eczema, a noncontagious inflammatory skin disease that can result in lesions on the feet depending on severity [3]. He was likely diagnosed with atopic eczema, the most common type of the disease, which often occurs in children on various parts of the body, including the feet. Atopic eczema is commonly treated with creams such as topical steroids which can reduce symptoms if applied twice daily [4]. If Power's cream was laced with botulinum toxin and applied to areas affected by eczema, the toxin would be absorbed through the lesions in his skin.

According to Sherlock Holmes, Moriarty introduced botulinum toxin, perhaps obtained from contaminated foods, into Powers' eczema cream. A study examining the hazards of botulinum toxin in foods found that cheese inoculated with *Clostridium botulinum* and incubated for 26 days contained up to 200 units/mL [5]. One unit of botulinum toxin is equal to 0.4×10^{-9} g of the toxin [6]. Therefore:

$$Poison\ collected = 200 \times 0.4 = 80 ng/mL = 0.08 \mu g/mL$$

Suppose that the top 5 mL of the eczema cream was laced with 5 mL of toxin with a concentration of 0.08 µg/mL, then the concentration of poison in the eczema cream would be diluted by a factor of 2, resulting in a final concentration of 0.04 µL. Furthermore, assuming that 1.5 mL of cream will cover the feet and will be applied twice daily with 80% absorption:

$$Absorption/time = (Toxin\ in\ cream \times (Cream\ applied)/day) \times percent\ absorbed$$

$$Absorption/time = 0.04\ (\mu g)/mL \times 3mL/day \times 0.8 = 0.1 \mu g/day$$

Powers would have been absorbing 0.1 μg of botulinum toxin per day, or 0.05 μg every time he applied his eczema medication. The amount of botulinum toxin required to kill half of all exposed individuals, referred to as the median lethal dose, is estimated to be 0.1 μg for humans [1]. Therefore, only two applications of toxin-laced cream could have proven lethal for Powers. In addition, the effect of botulinum toxin peaks at four to seven days and can last for up to six months, so the effects of subsequent doses would have been cumulative [7,8].

Onset of Symptoms
Botulinum toxin acts on cholinergic synapses in the peripheral nervous system to induce paralysis, ultimately leading to death [1]. The neurotoxin requires 24-72 hours to disrupt synaptic activity, and therefore the first symptoms of botulism do not occur until at least a day after contact with the toxin [7]. Therefore, Holmes' deduction that the poison was administered two hours prior to Powers' death was incorrect. Moriarty could not have planned for the poison to take effect while Powers was in the pool, as the onset of paralysis can vary by up to 48 hours. However, Moriarty's timing may still have been strategic because the first symptoms of botulism, which include blurred vision and difficulty swallowing, could have been ignored in the excitement of the swimming competition.

Conclusion
These findings show that the murder of Carl Powers could have been executed by mixing botulinum toxin into his eczema cream, though the scenario could not have occurred exactly as described by Sherlock Holmes. Although botulinum toxin cannot be absorbed through intact skin, the skin on Powers' feet would have had lesions and open sores due to his eczema, which could have facilitated the introduction of botulinum toxin into his bloodstream. The amount of toxin absorbed in this way would be small, but the high toxicity of botulinum toxin means even a minute concentration of the toxin could induce paralysis. A predicted time frame of one to two days would have been necessary for an adequate amount of toxin to be absorbed, and the toxin would require an additional one to three days to take effect.

References

[1] Davis, L.E. *Botulism*. Current treatment options in neurology, 5, 1, 23–31 (2003) http://www.ncbi.nlm.nih.gov/pubmed/12521561
[2] Geiger, J. (1924) *The possible danger of absorption of toxin of B. botulinus through fresh wounds and from mucous surfaces*. American journal of public health 14, 4, 309–10 (1924) http://www.pubmedcentral.nih.gov/articlerender.fcgi?artid=1354824&tool=pmcentrez&rendertype=
[3] Ring, J., Przyyilla, B., and Ruzicka T. Handbook of Atopic Eczema (Springer), 2, 2-5
[4] Torley, D., Futamura, M., Williams, H.C., and Thomas, K.S. *What's new in atopic eczema? An analysis of systematic reviews published in 2010-11*, Clinical and Experimental Dermatology 38, 449-456 (2013) http://onlinelibrary.wiley.com/doi/10.1111/ced.12143/abstract
[5] Kautter, D.A., Lynt, R.K., Lilly, T. and Solomon, H.M. *Evaluation of the Botulism Hazard from Imitation Cheeses*. Journal of Food Science 46, 3,749–750 (1981) http://doi.wiley.com/10.1111/j.1365-2621.1981.tb15341.x
[6] Hoffman, R.O. and Helveston, E.M., 1986. *Botulinum in the treatment of adult motility disorders*. International ophthalmology clinics. 26, 4, 241–50 http://www.ncbi.nlm.nih.gov/pubmed/3804629
[7] Nigam, P.K. and Nigam, A. *Botulinum toxin*. Indian journal of dermatology. 55, 1, pp.8–14. (2010) http://www.pubmedcentral.nih.gov/articlerender.fcgi?artid=2856357&tool=pmcentrez&rendertype=

[8] Shoemaker, C.B. and Oyler, G.A. *Persistence of Botulinum neurotoxin inactivation of nerve function.* Current topics in microbiology and immunology, 364, 179–96 (2013)
http://www.pubmedcentral.nih.gov/articlerender.fcgi?artid=3888862&tool=pmcentrez&rendertype=

The Big Fat Lie About Burning Fat

Emma Butcher
Honours Integrated Science Program, McMaster University
09/04/2014

Abstract

Carrying excess fat is a serious health concern, and yet many do not understand the basic principles of fat metabolism. This article explores the rate and path by which fat is excreted from the body, how this relates to exercise habits, and shows that it is not possible to reach a state in which only fat is metabolised for energy.

Introduction

Obesity is a serious global health concern that lowers the life expectancy of individuals and puts strain on healthcare systems. Many people attempt to lose weight by using pre-set "fat-burning options' on cardio equipment, which tend to be low intensity workouts. This paper will look into the viability of burning large amounts of fat under these conditions. The term 'weight loss' is a misnomer, as most people are concerned with fat loss rather than weight in general. Fat does not simply disappear; it must be metabolised and have the waste products removed from the body. The most abundant fat molecule in the human body, palmitoyl-stearoyl-oleoyl-glycerol ($C_{55}H_{104}O_6$)[1], is mostly carbon by mass. Carbon cannot be excreted by the body in sweat or urine, and fecal matter mostly consists of undigested food. Rather, carbon from fat and other energy storing molecules is exhaled as carbon dioxide [2]. If the body were to run purely on fat oxidation for energy, then fat would provide nearly all of the carbon dioxide exhaled in each breath. Theoretically, it is possible to calculate the amount of breaths needed to excrete the carbon in 1kg of fat under these conditions.

Determining the Amount of Carbon Exhaled

The potential mechanism for complete metabolism of fat could be described by [2, 3]:

$$C_{55}H_{104}O_{6(aq)} + 78O_{2(aq)} \rightarrow 52H_2O_{(l)} + 55CO_{2(aq)} + 34MJ$$

Metabolism of fat involves a variety of intermediates, such as Acetyl CoA, and is more complex than the previous equation. This equation describes the overall process, and will be used for simplicity sake. The mass of carbon dioxide produced from 1kg of fat can be determined stoichiometrically.

$$moles = \frac{mass}{molar\ mass}$$

Molecule	$C_{55}H_{104}O_6$	CO_2
mass	1.000x10³g	2.810x10²g
molar mass	8.614x10²g/mol	4.401x10¹g/mol
moles	1.161x10⁻³ mol	6.385x10⁻² mol

Table 1: Molecular data used for palmitoyl-stearoyl-oleoyl-glycerol and carbon dioxide [4].

From this, the volume of carbon dioxide is calculated assuming standard pressure conditions and an internal body temperature of 37°C [5] (or 310.15K) by employing the ideal gas law.

$$V = \frac{nRT}{P}$$
$$= \frac{(6.39 \times 10^{-2}\,mol)(8.21 \times 10^{-2}\,L\,atm\,K^{-1}mol^{-1})(310.15K)}{1\,atm}$$
$$= 1.63L$$

Determining the Number of Breaths

The average breath of a person at rest only uses part of the lungs' total capacity. This partial capacity is known as the tidal volume, and is typically 7ml/kg [6].

For the purposes of this example we will assume our subject has a mass of 70 kg.

$$Tidal\ volume = (7ml\,kg^{-1})(70kg) = 490ml$$

The average human expiratory breath contains about 5% carbon dioxide, while an inspiratory breath contains about 0.05%. This gives a net 4.95% of carbon dioxide expelled with each breath [7]. We can then use the tidal volume to calculate the volume of carbon dioxide in each breath.

$$V_{CO_2} = (490ml)(0.0495) = 24.3ml$$

In order to exhale the carbon dioxide formed in the metabolism of 1kg of fat (i.e. 1.63L of CO_2) they would need to take just over 67 breaths.

This calculation assumes that the person is burning fat while at rest. Burning fat at rest is caused by having a net deficit of calories, either as a result of diet or exercise [8]. However, if the subject is burning fat while exercising they would not be exhaling carbon dioxide in at the tidal volume. Rather they would be exhaling the expiratory reserve volume. Using our 70kg subject, we can predict their expiratory reserve volume to be about 1.0L [9], which is approximately double their normal tidal volume. Using the previously demonstrated method, we find that the subject will exhale about 49.5ml of carbon dioxide per breath. To expire the 1.63L of carbon dioxide from 1kg of fat would require a little less than 33 breaths. The exact number of breaths required would vary based on the subject's mass, sex, age, cardiovascular fitness, and health [10, 11].

Discussion

These calculations show it is theoretically possible to exhale the carbon of a kilogram of fat in only 33 to 67 breaths. However, this is obviously impossible in real life. If all exhaled carbon dioxide was the waste product of fat oxidation, then the body would have to be metabolising fat at a ridiculously rapid pace in order to produce 1.63L of carbon dioxide in this short time period. Thus, the body does not rely purely on fat oxidation, and so there must be other processes contributing to the carbon dioxide. The majority of carbon dioxide produced by the body comes from waste products of metabolism of glucose and glycogen for energy. Exhaled carbon dioxide comes from a combination of these processes at any given time [12].

Evidence shows that the body uses a higher percentage of fat in 'the fat-burning zone', which is reached by doing low intensity exercise at 55% to 65% of maximum heart rate [13]. Many models of cardio exercise equipment have 'fat-burning' settings that help the exerciser stay in this range. Unfortunately, there is a common misconception in the public that the body only uses fat for energy in this zone. Although the *percentage* of fat oxidation in this zone is higher than at other ranges, the total amount of energy used is less than in higher intensity exercise. This results in less fat loss over the same period of time [13]. For example, brisk walking (5.5km/h) may use a higher percentage of fat for energy than jogging (10km/h), but walking for half an hour only burns about 130 calories, while jogging would burn about 400 calories [14]. To overcome the fatigue brought on by high intensity exercise, interval training can be used. By taking short breaks between bouts of exercise the person can reoxygenate and reduce lactic acid build up, but still maintain a high heart rate for long periods of time [15].

The previous calculation supports this evidence. Carbon dioxide concentration in breath does not significantly change during exercise [7], so if all exhaled carbon dioxide was the waste product of fat oxidation, we would expect to see results similar to the previous example, in which the subject exhaled a kilogram of fat in 33 breaths. This would indicate that fat loss was occurring at a ridiculously rapid pace, and so it is clear that the body does not resort to metabolising only fat during exercise.

Conclusion

It is unreasonable to assume that the body can reach a state of using only fat for energy under normal conditions. Rather, fat oxidation works in conjunction with other metabolic processes. Although low intensity exercise does burn fat, high intensity workouts, particularly those that employ interval training, are expected to be more effective in causing fat loss and improving overall health.

References

[1] Madihally, S. V., 2010, *Principles of Biomedical Engineering* (Artech House).
[2] Marieb, E.N.,2006, *Essentials of Human Anatomy and Physiology* (Pearson Education Inc.), 8thed, Ch 14.
[3] Frayn, K.N, 1996, *Metabolic Regulation: A Human Perspective* (Blackwell Publishing), 2nd ed,Ch 11.
[4] Dayah, M., 2013, *Ptable* [online]. Available at: http://www.ptable.com/ [Accessed 3 March 2014].
[5] Wilkinson, R.T. et al., 1964. Psychological and Physiological Reponses to Raised Body Temperature. Journal of applied physiology, 19(2), pp.287–91.
[6] Beardsell I, Bell S, Robinson S., and Rumbold H., 2009, *Get through MCEM part A: MCQ's* (Taylor and Francis group).
[7] Taylor, C., Lillis, C., and LeMone, P., 1989, *Fundamentals of Nursing (J. B. Lippincott Company).*
[8] *Edward, C.F., 1995, Fat Metabolism During Excersize, Sports Science Exchange, vol 8, number 2.*
[9] *n.a., 2007, The American Heritage Medical Dictionary(Houghton Mifflin).*
[10] Donnelly, P., Yang, T.-S., Peat, J., Woolcock, A.,1991, *What factors explain racial differences in lung volumes*, European Respiratory Journal, 4, 829-836.
[11] Patrick, J.M. and Howard, A., 1972, *The influence of age, sex, body size and lung size on the control and pattern of breathing during CO2 inhalation in Caucasians*. Respiration Physiology, 16(3), pp.337–350
[12] Blake, J., Munoz, K., Vople, S., 1995, *Nutrition: From Science to You* (Pearson Education Inc.), Ch 8.
[13] Leyland, T., 2007. *The Myth of the Fat-Burning Zone*. The Crossfit Journal Articles, (54).
[14] My Fitness Pal, Calories Burned From Exercise [online]. Available at: http://www.myfitnesspal.com/exercise/lookup
[15] Laursen, P.B. & Jenkins, D.G., 2002. *The Scientific Basis for High-Intensity Interval Training*. Sports Medicine, 32(1), pp.53–73.

Journal of Interdisciplinary Science Topics

Calculating the Punching Power of "One-Punch" Mickey

Daim Sardar
Honours Integrated Science Program, McMaster University
09/04/2014

Abstract
This paper will examine the film *Snatch*, to understand concepts related to the physics of a knockout in boxing. The punching force exerted by the character named Mickey is calculated by using the equations for momentum and impulse (under reasonable assumptions) to find that Mickey exerts a punching force of 1780N.

Introduction
The movie *Snatch* is a British film directed by filmmaker Guy Ritchie, which takes place in London, England. One of the characters in the film is depicted to have supreme fighting skills as he can knockout any individual with just one punch. This character is named Mickey O'Neil, also known as "one-punch" Mickey, portrayed by American actor Brad Pitt. The movie shows several scenes where Mickey finishes his opponent via knockout in a boxing match with just one punch.

As such, the premise for this article is to analyse the effects of a concussion on the brain and what creates the "knockout" effect on an individual. The article will also investigate the punching force exerted by Mickey by applying the laws of physics to determine the force behind his knockout power.

Mechanism of a Concussion
The term concussion is defined as a sharp blow to the head in the form of a collision [1]. More specifically, in boxing this refers to the impact of the glove of one person to the head of the other creating a relative motion of the head and brain, adhering to Newton's laws of motion. Within the skull, the brain is suspended between cerebrospinal fluid and attached by blood vessels and nerve fibres [1]. During the impact, the head is accelerated due to inertia that causes the skull to accelerate as well. As the skull comes to rest, the brain is still moving within and it will hit the wall of the skull, creating a contusion. If punched hard enough, the reticular activating centre (important in maintaining proper posture) in the brain will be affected leading to an irregular reflex which will cause the person to fall and hit the floor unintentionally [1].

Applying Physics to Boxing
To calculate the punching force exerted by Mickey to knockout his opponent, concepts from physics can be used. When a boxer is in their "ready" position, their fists are up and contain potential energy that will come from their muscles and body motions. During the punching process, the potential energy is being converted to kinetic energy.

In order to calculate the punching force, we need to calculate the change in momentum for the head being punched. We know that before the punch, the momentum of the head will be zero; therefore the change in momentum will equal the final momentum (head being punched). A few reasonable assumptions have to be made however in order to accommodate the calculations:

- The weight of the head of the person being punched by Mickey in the movie is 8kg [3]
- The time of contact between the hand and head is 0.03s [4]
- The head moves a distance of 0.20m during contact as estimated by watching the film

The first step is to calculate momentum given by the following equation [2];

$$\rho = mv$$

In the equation above, velocity can be calculated by [2];

$$v = \frac{d}{t} = \frac{0.20m}{0.03s} = 6.67 \, ms^{-1}$$

Substituting this value into the momentum equation will give us;

$$\rho = mv = 8kg \times 6.67 ms^{-1} = 53.4 \, kg \, ms^{-1}$$

Since we now know the change in momentum, we can use the equation for impulse to find the force since the change in momentum is equal to the impulse. The equation for impulse is the following [2];

$$Impulse = F \times \Delta t$$

Rearrange for 'F' to get;

$$F = \frac{Impulse}{\Delta t}$$

The final step is to substitute the values into the rearranged equation for 'F' and find the force. The change in time is 0.03s as stated in the assumptions.

$$F = \frac{53.4 \, kg \, ms^{-1}}{0.03 \, s} = 1780 \, N$$

Conclusion

By conducting calculations using the assumed values, it is seen that the force exerted by Mickey's punch is 1780N, which is about 400lbs of force. Experimentally obtained values from professional boxers in the middleweight class (165lb) have been observed to be around 2625N of force which is fairly close to "one-punch" Mickey's (given that he can be comparable to a middleweight due to his size in the movie) [4]. The calculations however, cannot take into account the way an individual places a punch on the opponent's head, which is an important aspect of achieving a knockout. Delivering a punch in a way that creates a "snapping" motion (e.g. landing a hook punch onto the jaw) in the opponents head will have a higher chance of knocking the opponent out because it will cause the brain to move faster and hit the wall of the skull, causing a concussion. The 1780N of force calculated here is a reasonable calculation of Mickey's punch, but it is important to mention that a great deal of his destructive power is related to the placement and style of his punch.

References

[1] Allan, R.J., *Protecting the sportsman's brain (concussion in sport)*, British Journal of Sports Medicine **25(2)**, 81 (1991).
[2] Knight, R.D., (2008). *Physics For Scientists And Engineers: A Strategic Approach* (Adam Black), 2nd edition. Pearson Education.
[3] Plagenhoef, S., Evans, F.G. and Abdelnour, T, "Body Segment Data" [Online]. Available: http://www.exrx.net/Kinesiology/Segments.html. [Accessed 13 March 2014].
[4] Walilko, T.J., Viano, D.C., & Bir, C.A., *Biomechanics of the head for Olympic boxer punches to the face*, British Journal of Sports Medicine **39,** (2005).

Journal of Interdisciplinary Science Topics

Golden Fleece: An Ancient Sheep

Sean Gilmore
The Centre for Interdisciplinary Science, University of Leicester
09/04/2013

Abstract
Of the many infeasible creatures and relics of ancient Greek mythos, the Golden Fleece from Jason and the Argonauts has drawn much attention from historians as to what it represented in terms of politics, technology and religion. However, we will instead explore the scientific basis to the possibilities of the existence of a gold fleeced ram. This article specifically addresses the biological and geochemical aspects to this multidisciplinary problem, and follows previous research carried out by this group regarding the physical and biophysical aspects.

Introduction
Written in the 3rd century BC, *Argonautica*, by Apollonius Rhodius, is one of the oldest and most complete accounts of Jason and the Argonauts, a tale that is one of the most famous and foundational in Greek classical history. The agonist of *Argonautica*, one Jason of Iolcos, quests across uncharted seas and overcomes impossible obstacles in order to obtain a legendary relic, with promise of his rightful crown. This relic, the fleece of a ram with wool of gold, can almost be described as unassuming among the many fanciful creatures and relics in the tales of ancient Greece. It is the fleece's representation of power, rather than any monetary value, which is at the heart of Jason's quest. However, a fleece of gold must have once belonged to a creature capable of creating its fleece as a product of gold rather than the usual keratin.

In slightly more recent history, in 1953, the Russian chemist Vinogradov published an extensive work concluding that the accumulation of gold by organisms was entirely random and that there was no evidence of auriferous animals in any part of the world [1]. However, if such a creature were to exist, it would raise a number of interesting issues regarding physical, geochemical, biochemical, and evolutionary feasibility.

Old Sheep
Work previously done by this group showed that assuming a fleece with the same proportions as that of a normal sheep's wool fleece, the Golden Fleece would weigh approximately 50kg, almost the weight of another entire ram on its back. Vinogradov's statement is seen to be correct on the most part; the maximum gold content in marine invertebrates is found to range from orders of 0.001 to 0.1 nanograms per gram (ng/g) of dry biomass, while the largest measurements range from 2 to 50 ng/g for migratory birds collected in an auriferous area [2].

Assuming the body water content of the ordinary 66kg ram is roughly 60% of its full bodyweight, the dry biomass of a 116kg golden fleeced ram will be constituted of over 65% gold. This is 130 million times that seen in the most aurous of known terrestrial vertebrates [2]. No creature has so far seen to have such large proportions of any metal. There are, however, examples of bacteria that utilise metal nanoparticles, and a recently discovered deep-sea vent-dwelling snail with iron plated foot scales. This suggests the bio-capability of producing metal constituent body parts in simple organisms, yet it would be a significant extrapolation to extend this to the vertebrates. Much less likely is the specific case of gold. This partly due to being among the least reactive metals, and therefore, gold is difficult to implement into biological systems. More notably, gold is such a rare metal that a species is unlikely to have a reliable enough source in order to evolve a use for it, and an organism that can is unlikely to stumble upon enough to make any use of it.

For example, the gold content of water sources ranges widely, with sea and river water from non-auriferous regions having gold content per litre measuring below the nanograms [2], while river and ground water from auriferous regions has been found to reach respective gold concentrations of 15.44 and 67.82 ng/L.

If we give the ram the benefit of the doubt, and we assume it obtained its gold from groundwater equivalent sources in auriferous regions, it would still have to consume at least 750 billion litres of water to grow its 50kg coat. Using data from the Ministry of Environment [3], a ram consuming the recommended

11 litres per day would take 180 million years to do so with the anatomical capabilities of an ordinary sheep. This assumes all gold is used for fleece growth and not subsequently damaged, lost or eroded; an unlikely prospect due to available weathering time and the natural malting of the primitive breeds on which the ram is modelled.

Cold Sheep
Another constraint relates to the large difference in the thermal properties of wool and gold. The thermal conductivities of bulk steel and keratin are known, as are those of steel wool and sheep's wool. The thermal conductivity of gold is 5 to 10 times more than that of steel [4, 5], and 500 times that of bulk horn keratin [6].

Using the properties of both sheep's wool [5] and steel wool [7] to model the aurous wool of the ram, the fleece is likely to have thermal conductivity of 10 to 40 $Wm^{-1}K^{-1}$. This is 250 to 1000 times more conductive than ordinary sheep's wool. Therefore, as a warm blooded vertebrate, the sheep would overheat easily in the sun, and require a far thicker coat than the model rams in order to maintain body heat in the harsh Caucasian mountains. This causes a long cascade of further issues due to increased weight and the subsequent gold requirement of the ram. Although the evolution of such the creature is already infeasible, these further issues point clearly and evidently towards the likelihood that a ram with golden wool could not exist outside of the ancient Greek mythos.

References

[1] A. P. Vinogradov, The Elementary Chemical Composition of Marine Organisms, New Haven: Sears Foundation, 1953.

[2] R. R. Brooks and R. W. Boyle, "Chapter 7 - Animals and Noble Metals," in Noble Metals and Biological Systems: Their Role in Medicine, Mineral Exploration, and the Environment, R. R. Brooks, Ed., CRC Press, 1992, pp. 179-215.

[3] Environment and Resource Division (Water Management Branch), "Animal Weights and their Food and Water Requirements," Ministry of Environment, Lands and Parks, 2001. [Online]. Available: http://www.env.gov.bc.ca/wat/wq/reference/foodandwater.html#table1. [Accessed 18 March 2014].

[4] The Engineering ToolBox, "Thermal Conductivity of Metals," [Online]. Available: http://www.engineeringtoolbox.com/thermal-conductivity-metals-d_858.html. [Accessed 17 March 2014].

[5] The Engineering Toolbox, "Thermal Conductivity of some common Materials and Gases," [Online]. Available: http://www.engineeringtoolbox.com/thermal-conductivity-d_429.html. [Accessed 18 March 2014].

[6] S. Baxter, "The thermal conductivity of textiles," Proceedings of the Physical Society, vol. 58, no. 1, p. 105, 1946.

[7] K. J. Russell, "Are Good Electrical Insulators Also Good Thermal Insulators? A Study of Thermal Conductivity," University of Southern California, Los Angeles, 2006

www.ingramcontent.com/pod-product-compliance
Lightning Source LLC
Chambersburg PA
CBHW081050170526
45158CB00006B/1924